The Plant Detective's Manual:

a research-led approach for teaching plant science

The Plant Detective's Manual:

a research-led approach for teaching plant science

Gonzalo M. Estavillo, Ulrike Mathesius, Michael Djordjevic and
Adrienne B. Nicotra

Australian
National
University

eTEXT

ANU
eTEXT

Published by ANU eTEXT

The Australian National University

Canberra ACT 0200, Australia

Email: etext@anu.edu.au

This title is also available online at http://press.anu.edu.au

National Library of Australia Cataloguing-in-Publication entry

Creator:	Estavillo, Gonzalo, author.
Title:	The plant detective's manual: a research-led approach for teaching plant science / Gonzalo M. Estavillo, Ulrike Mathesius, Michael Djordjevic and Adrienne Nicotra.
ISBN:	9781925022179 (paperback) 9781925022186 (ebook)
Subjects:	Botany -- Research.
	Plants -- Analysis.
Other Creators/Contributors:	
	Mathesius, Ulrike, author.
	Djordjevic, Michael, author.
	Nicotra, Adrienne, author
Dewey Number:	580.7

Cover design and layout by Cariboo Design.

Preface for The Plant Detective's Manual

Science's contribution to revolutionising agriculture is well-known. In particular, chemistry, biology and genetics have allowed us to produce more and more, with less and less.

Now we face unprecedented global challenges – climate change, an estimated 925 million people without adequate nutrition, and an ageing farming workforce.[i]

On top of these, the world's population is predicted to increase to 9.2 billion by 2050, requiring an increase in global food production of 70 per cent.[ii]

Providing enough food in this context will be an unprecedented scientific, economic and political challenge.

It is through the scientific method and evidence-based reasoning that our scientists' capacity to respond to these challenges will grow.

The plant biologists and agriculturalists of tomorrow will learn from the experienced researchers of today.

I have long been advocating curiosity-driven and problem-based science teaching – science as it is practised – alongside the subject-specific knowledge that science requires. *The Plant Detective's Manual* does just this.

Led on a journey of inquiry, students are presented with real-life problems. Students form hypotheses, design experiments and evaluate the collected evidence.

In learning science as it is practised, students will be armed with the critical-thinking skills needed to face the challenges of tomorrow.

I commend the *The Plant Detective's Manual* to teaching science and wish you well in your future endeavours.

Professor Ian Chubb AC
Chief Scientist for Australia
December 2014

---

[i] World Hunger Education Service (2012) *2012 World Hunger and Poverty Facts and Statistics.*
http://www.worldhunger.org/articles/Learn/world%20hunger%20facts%202002.htm
[ii] Croplife (2012) Submission in Response to National Food Plan Green Paper. Introduction, paragraph 4.
http://www.croplifeaustralia.org.au/files/newsinfo/submissions/2012/CropLife%20Submission-National%20Food%20Plan.pdf

Contents

The inspiration for the Plant Detectives Project

If global challenges in food production and the impact of ever-declining biodiversity are to be tackled, every country will need plant biologists who have a deep understanding of plant morphology, physiology and genetics, and how these interact to affect plant function in changing environments. These scientists will also need the capacity to use an effective and powerful set of technologies and research strategies.

These were the motivations behind our redesign of a second-year undergraduate plant science course at The Australian National University in 2007. 'Plants: Genes to Environment' ran successfully for many years, but we wanted to do more to inspire students and instill a meaningful involvement in the practical classes. We wanted our students to grapple with the scientific method, to learn at firsthand about the inquiry process, hypothesis development and analysis and interpretation of evidence.

To do this, we created an integrated set of laboratory investigations that we felt truly reflected the mysteries of plant biology and puzzle-solving processes that we had encountered in our research experience. Rather than a set of unconnected experimental activities, we created a series of closely related experiments that focused on solving 'mysteries' in the life of the plant *Arabidopsis thaliana* (thale cress). The activities charge students with finding the 'suspect' gene responsible for the specific phenotypes of an unknown Arabidopsis mutant, which are encountered when they expose the plants to different environmental stresses. This, we hoped, would give keen but inexperienced student scientists a realistic taste of the joys (and frustrations!) of plant science research.

We chose Arabidopsis because of the extensive collection of genetically mapped mutants that are readily available. We challenge budding plant biologists to resolve a biological puzzle, namely the mystery of 'unknown' genetic mutations affecting plant form and function. As they do this, students learn to apply the basic concepts in plant biology that they have learned in lectures and from their course readings in a research context. This Plant Detectives Project manual is the tried and tested outcome of several years' experience guiding our students through authentic plant science experiments that help them become astute researchers and 'plant detectives'.

In our teaching we colour the class with this inquiry-based approach. First, we partially 'flip' the classroom: students are given pre-lecture readings and discussion questions. The students' answers to these questions, which are presented in group discussions at the beginning of each theory class, drive the lecture format, with the teaching staff emphasising the aspects of that day's material least understood by students. With this format, we have been able to compress the theory portion of the course into the first half our teaching semester (13 weeks). This means that by the time the laboratory activities begin — a third of the way into the course — the students have a solid theoretical background under their belts. This is tested through an examination that is given two-thirds of the way through the semester. Thus, in the last weeks of the course the students are entirely devoted to the research element of their work.

As is common in practical science work, students conduct their plant detectives projects in groups of three to five, and work intensively within those groups. Our second innovation, therefore, was to enhance collaborative engagement and break down the barriers between those groups by instituting cross-group 'lab meetings' at the start of each practical. These meetings consist of groups of about five students that include one member from each of several project groups. Students share their group's results from the previous week, and compare the hypotheses and objectives for the current week. These collaborative lab meetings give the students practice in communicating their results, reflecting on their discoveries, and taking new ideas back to their project groups. This mirrors the collaboration of plant science researchers across the world.

Finally, we conclude each course with a symposium session in which each group presents their results to the class. These sessions are rewarding for students and teaching staff alike, as together we see how the students make links between gene, structure, and function in the environment.

The assessment for the course is a combination of a small number of marks for participation in the discussions in lecture, the symposium presentation, the exam and a final report in the format of a scientific paper. Together, these elements enhance our students' learning environment and the value of the Plant Detectives Project as a learning experience.

Although thrilled by numerous university and national awards for our innovative teaching, we have been most excited by the interest in our ideas and experimental approaches from other plant science educators in Australia and overseas, who are also seeking to improve their plant biology curriculum and attract more students to plant sciences. We are thus proud to present this manual as a gift to our colleagues worldwide. Here you will find a detailed collection of state-of-the-art procedures in plant biology, as well as background information on more commonly used techniques, and tips for class preparation. The concepts and methods we present can be adapted to meet the specific needs and expertise of the teaching staff, and provide inspiration for scaling up for larger audiences, or simplifying for more junior classes. Through this publication, we hope to support our teaching colleagues in making a significant impact on improving the learning experience of plant biology students worldwide, and hope that we will motivate and inspire a new generation of plant detectives.

I learned more from the labs in this course than all the other labs from my other courses combined.

One of the most interesting and influential courses I have taken in my degree.

I loved [this course] and have already recommended it to my first-year friends.

The different style to other courses was educational and refreshing. The experimental work was particularly stimulating and challenging. It gave me a glimpse into what real research would be like and a feel for some of the practicalities and difficulties.

It was really hard and challenging for me at the beginning to speak out during lectures and to have confidence in myself—I [couldn't] concentrate on the lecture as I was too nervous thinking of what I should say if the lecturer suddenly pointed at me - but as time passed I didn't even realize how much my attitude towards learning (and life) changed for the better.

[It was] interesting going from the genetic/molecular level to a physiological level.

The symposium was a surprisingly relaxed exchange of ideas. It felt like the markers were genuinely interested in the presentations rather than simply having to be there to assign a grade, and seeing all the other groups' work helped me think about what my group had done in clearer terms. ... After the talks was a good opportunity to pick the lecturers' brains, and also have a more relaxed chat as the teaching component of the course reached its end. ... It'll be a long time before I could face counting any part of an Arabidopsis plant you'd care to mention, but hey, I ended up with a brand new scientific skill set ... (Extract from student blog).

About this manual

This manual provides a collection of detailed protocols to perform plant physiology experiments. Our aim was to design and implement a series of related experiments for you, the student, to experience a real research project. We expect that these activities will help you to acquire basic experimental techniques and skills in results interpretation. The techniques are a compilation of research methods that we have established over the years. The type of experiments and research questions used in this manual are based on our experience in studying the model plant Arabidopsis as well as in other species. Our project will focus on Arabidopsis because of its suitability to the research approach and because of the breadth of resources available.

The manual starts with a general information section describing the experimental system, main objectives and expected outcomes. Next, we describe an 'Outline' of activities (or experiments) which detail procedures to perform a variety of experiments, ranging from morphological analyses to investigation of photosynthetic parameters.

Each activity is divided into the following sections:

Introduction and objectives: a brief introduction presenting topical information to put the activity into context and outlining the main objectives.

Materials: a list of materials (both chemical solutions and instruments) required for the activity. Warning signs using the hazard pictograms from the Globally Harmonized System of Classification and Labelling of Chemicals are used to denote chemicals that are hazardous. We recommend you read Appendix A: General rules for safety and conduct, for additional information.

Procedure: a detailed explanation of the experimental procedure follows in table format. The right column describes the procedure while the left column indicates which materials are required for specific steps. All links to external websites are also described in text format in Appendix D: Databases and web resources. 'Alternative' procedures have been included with the aim to provide other (usually less costly) ways to run the experiments. 'Optional' refers to activities, or part of an activity, that are may not be required; consult your instructor.

Expected outcomes: individual expected outcomes for each activity are also detailed to help you to understand the kind of data that has to be collected or how the data should be analysed and the results presented.

The last activity subsection describes the expected outcomes and data to be recorded. Activity 12 provides directions about writing a scientific manuscript based on the instructions for contributing authors of the peer-reviewed journal *Functional Plant Biology*. We also provide basic tips and advice on grammar and style for writing a scientific paper as well as for the oral symposium presentation.

We have included a section introducing you to principles of experimental design and the text boxes at the start of each activity indicate which of these design principles are more relevant for that activity. Appendix B: What do I do with my data?, is essential reading for the early career plant detective.

General information about the Plant Detectives Project

Introduction

The rate at which we must increase food production needs to grow with every increase in world population growth (United Nations Food and Agriculture Organisation; http://www.fao.org/docrep/x0262e/x0262e23.htm). This scenario is further complicated by the uncertainty of climate change and biodiversity loss. Meeting increased food demand under ever limiting conditions while protecting natural resources will be a major challenge. The next generation of plant scientists will need creativity backed by high quality knowledge and investigative skills if they are to tackle this challenge.

We believe that a research-led approach provides an excellent theoretical and practical understanding of the integral links between genes, cells, whole plants and the environment that will be required for solving some of these future challenges. Over the next weeks, you will put into practice your newly acquired theoretical knowledge as you apply cutting-edge laboratory techniques to the "puzzle" of pinpointing genetic mutations in plants. In this process, you will get hands-on experience, test essential concepts in plant biology and explore, investigate and assess the effects of genetic mutations on plant form and function. We hope this will motivate and inspire you to learn more about plant biology, and help you develop the problem-solving curiosity, questioning and resourcefulness that one needs to be an effective researcher.

Mutants play a major role in understanding how plants work

The isolation and identification of mutant plants has played a major role in understanding plant physiology and is a regular screening procedure for genetic studies. Although mutations can arise naturally in all species, it is relatively easy to induce changes in a plant's genetic makeup by treatment with chemicals. Observation of the resulting external and internal features, physiological changes and molecular profiles (i.e., transcripts, proteins, and metabolites) caused by a mutation can provide important information about the role of the gene that is affected.

This laboratory manual provides laboratory techniques for the study of plant morphology and physiology of the model organism thale cress (*Arabidopsis thaliana*, hereafter Arabidopsis). The set of activities used here is not, however, limited in application to this well know model system; they can be applied to any species. It is hoped that the laboratory exercises presented, along with the appropriate theoretical information, will provide you with fundamental knowledge of the links between genes, plant form and function.

Why did we choose Arabidopsis?

The flowering plant Arabidopsis is a member of the mustard family (*Brassicaceae*) native to the Northern hemisphere (Fig. 1A-B). Arabidopsis has been extensively used as a model plant for genetic and physiological studies for half a century (Meinke *et al*. 1998) and has played a major role in our understanding of plant biology (Lavagi *et al*. 2012). Arabidopsis' small diploid genome (125 Mb) was the first plant genome fully sequenced (Initiative 2000). Arabidopsis has five chromosomes and a short life cycle (six weeks from germination to mature seed) and its small physical size makes it particularly amenable for laboratory research. There are over 7000 natural accessions, or "ecotypes", of Arabidopsis, each with distinctive genetic make ups providing adaptation to specific environmental conditions (Weigel 2012). There is also a plethora of mutant lines of Arabidopsis, some natural and some generated in the lab. Both ecotypes and large mutant populations are available through stock centres, like Nottingham Arabidopsis Stock Centre (NASC) and the Arabidopsis Biological Resource Centre (ABRC) in Ohio.

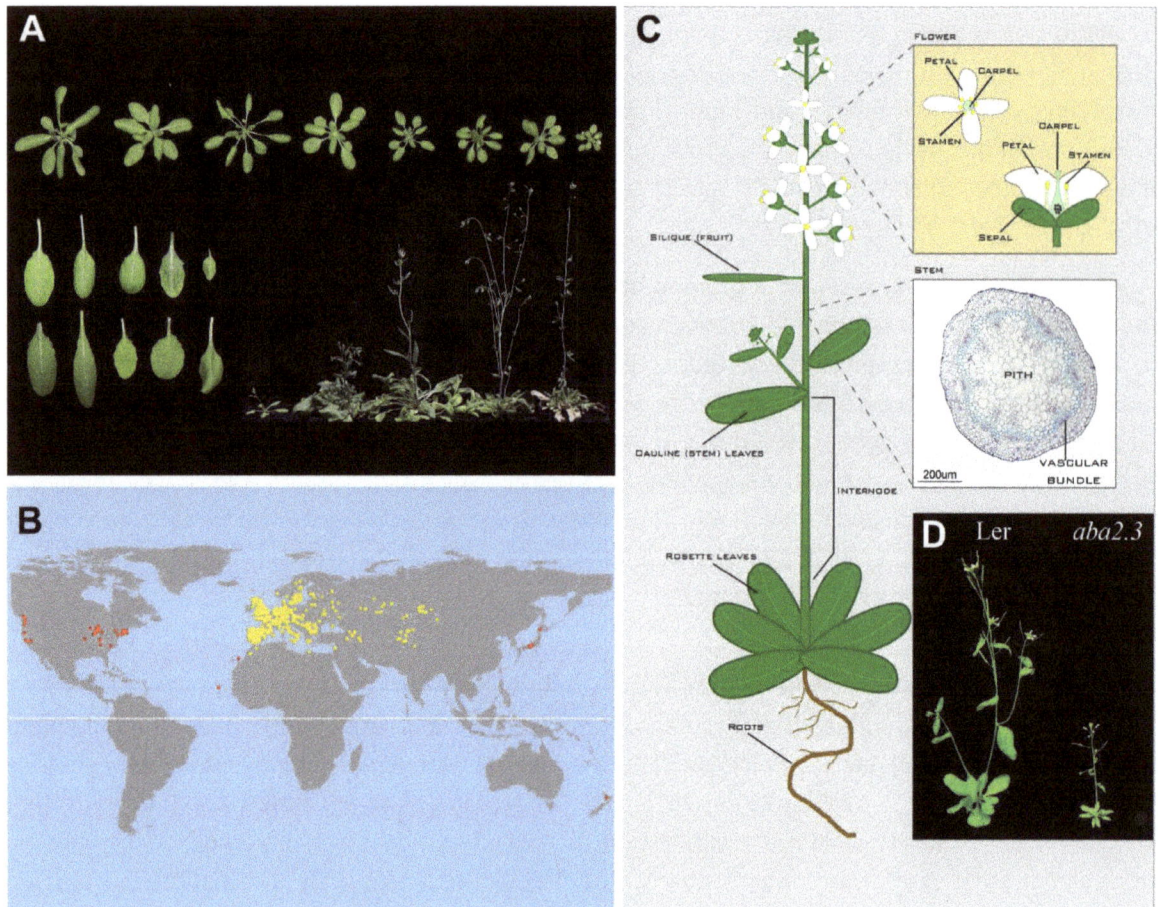

Figure 1. Arabidopsis accessions, distribution and morphology.
A) Top and side views, and individual leaf number 6 of different Arabidopsis accessions after 4 weeks. B) Distribution map of known accessions. Red dots represent likely introductions. C) Schematic representation of Arabidopsis organs. D) Example of Arabidopsis mutant. A, B, adapted from (Weigel 2012).

One way of generating Arabidopsis mutants is by treating seeds with chemicals, such as ethyl methanolsulfonate (EMS), or radiation to create lesions in the gene sequence. Seeds are sown and the resulting M1 generation grown to flowering stage. At this stage, it is expected that any particular lesion in a DNA locus would have occurred in only one of the five chromosomes. M1 individuals are "heterozygous" for that particular mutant allele. The M1 plants produce flowers, and because Arabidopsis plants self-pollinate (meaning that the pollen and ovule come from the same individual) they will go on to produce seeds. Some M1 plants harbouring dominant mutations can present in a different *phenotype* (that is, observable characteristics). To detect the effect of recessive mutations, the self pollinated seeds from the M1 plants are sown, giving rise to a homozygous M2 generation. Seeds from the M2 generation are then collected and, either the seeds or the resulting plants can be screened for recessive mutations by comparing specific traits to the wild type, such as morphology, pigmentation, etc. (Fig. 1D). It is expected that 1 out 4 plants will be homozygous for specific mutant alleles in the M2 plants. Because random mutations can occur in other loci, the isolated mutant plants are usually backcrossed to the parents in order to "clean" any spurious secondary mutations.

Working with Arabidopsis

Arabidopsis plants are commonly grown either on plates containing nutrient medium and agar, or in soil. Soil can be supplemented with slow release fertilizers, such as Osmocote Mini Exact, or Hoagland's solution. Because of their small size and short life cycle, they can be grown in high density in either growth chambers or greenhouses. A major factor in plant morphology and life cycle is the photoperiod (i.e. hours with light within a 24 hour period). Plants grown at shorter photoperiods (e.g. 8 hours light) grow slower and take longer to flower than plants grown at longer hours in the light (e.g., 16 hours). Watering by subirrigation is preferred and the amount of water and frequency of watering depends on the specific pot size and number of plants. TIP: It usually takes ~800 ml of water per 31 x 44 cm tray every 2-3 days to keep plants growing happily. It is very important not to overwater: if the top soil is wet, do not provide additional water.

Objectives of the Plant Detectives Project

The major objective of the Plant Detectives Project is to demonstrate the links between variation at the genetic, molecular and phenotypic level in different types of environments. In this case, our experimental setting will be a glasshouse, or growth chamber, and you will subject the plants to a range of growth conditions (e.g. drought) to assess the impact of genetic changes on the plant.

In this project you are a "plant detective" who needs to use the scientific method to discover genetic changes manifested by differential responses between the mutant and the wild type. For this, you will be given seeds from wild type and mutant Arabidopsis. Although unknown to you, the mutant will be one that has **already been published**[1]. Only one member of the teaching team will know which mutants are being used or which students have them. Your demonstrators, peer mentors and academics will not know the identity. In each practical session, you will perform a set of experiments to identify the phenotypic effects of the mutation on the plant morphology, anatomy, physiology and biochemistry. It is expected that at the end of the class you will be able to infer your "unknown" mutant by comparing your findings with those in the literature.

By doing this, you will become familiar with some of the most widely used and important techniques in contemporary plant biology around the world. You will also learn about plant physiology and morphology by monitoring the changes in the wild type and mutant plants in response to different environments. These techniques are regularly used by plant biologists (including geneticists, ecologists, physiologists, biochemists, developmental and molecular biologists). They are the current standard, and you will find they have very broad application.

To meet the above objective you will work in groups and will investigate your unknown Arabidopsis mutant by performing the following assays:

1. Calculating seed germination rates, root growth and development
2. Observing internal and external phenotypic features throughout germination and development
3. Analysing pigment composition of plant parts
4. Measuring the gas exchange characteristics (photosynthesis and transpiration)
5. Assessing effects of water availability (and potentially other growth treatments of your choosing) on the above plant characteristics

---

[1] Note to instructors: the Plant Detectives web site includes a repository of recommended mutants that have been used to good effect in the past. We welcome additions to this list if you find other mutants that work well.

Before the practical component

You will need to complete a brief on-line quiz before each laboratory session. This is to encourage reading of the appropriate protocol before you come to the lab so that you know in advance what is expected of you and what the day's objectives are. You will receive 2% for completing each quiz. Quizzes will be available online for 1 week before the laboratory session, and will close 15 minutes before the session begins.

Please note, these Quizzes are largely going to run on an honour system. You are welcome to discuss the answers among yourselves before completing the quiz. We will not give the 2% to any students whose answers are identical or for nonsense answers.

The questions on the quiz will be the same each week and are listed below. Read your lab protocols and your class notes in order to complete the quiz.

1. Describe and interpret your results from the previous lab session (from session 2 onwards)

Compile all the data (and share between your group members) and describe the results that you've obtained from the previous experiment. The results need to be presented in a conclusive form. Give us numbers or percentage, for example, "The germination rate for the mutant is 54%". We don't want vague results like "Some of the mutants germinated". The results will then need to be interpreted accordingly. You may refer to your lab manual, lecture notes or additional journals to interpret your results. See Appendix B for tips on how to handle your data. This will also help you with the write up of your final report.

2. Describe this week's experiment

Read your lab manual and understand the experiment that you'll be doing for this week. Understand the rationale behind the experiments and summarize about what you'll be doing in the lab for this week's prac.

3. Explain the objectives of this week's experiment

These are described in each activity, but we'd like you to put them in your own words. Explain what you understand about the objectives of the experiment. Dot points are fine but be specific.

4. Explain one of the techniques for this week's experiment in more detail

Some of the experiments will require you to use more than one technique. You may choose any of the techniques for this week's prac and explain what you understand about the technique, including the purpose and the applicability of the technique in regard to your experiment.

5. What is your hypothesis for this week's experiment in regard to your mutant?

Based on your growing understanding of your mutant (e.g. results from previous week's experiments) and the objectives and experimental techniques that you'll be doing this week, try to state a hypothesis regarding your mutant. Make your hypothesis precise and descriptive: not simply 'I hypothesize that x will be different from y', but 'I hypothesize that *x* will be greater than *y* because .....'

During the laboratory session: Take notes on the practical material in your laboratory notebook. Number the notebook pages and record the date during each class. We expect you to make notes of what you did, record observations and describe results, draw pictures of anatomical structures etc. We will provide questions to guide your thinking. Your entries will serve as notes to yourself. If you take your laboratory journal seriously you will end up with an excellent study tool and reference for future years. Record what you learned and keep notes on questions that you have about what you have done. Bring your notebook to every laboratory session

Expected outcomes

The Plant Detectives project has no "right or wrong" outcome. Instead, the goal is to use this model system to give you a chance to explore 'doing science'. You will have the opportunity to make your own links between gene (or genes) and phenotype. Based on your interpretation of results and literature research you will develop a solid understanding of modern practice and the growing connections among the fields of genetics, physiology, biochemistry and plant ecology.

Assessment

The goal of this project is to identify the mutant, i.e. identify the function of the mutated gene, or at least hypothesise what it could be, and to describe the effects of the mutation on the phenotype. This will be based on results of the lab work and literature research. The final mark is not based on whether or not the right answer (i.e., the gene) is identified, but on the approach taken, and the interpretation of the results used in identifying the likely suspects.

Assessment will consist of a write-up in the format of a scientific paper and a symposium presentation to the class. The class presentations in particular are a great opportunity to share with your peers the outcome of you research. Since each group is investigating a different mutant, the symposium will showcase a broad range of gene to environment connections.

Working groups and peer mentors

Students will be organized in "Teams" of two to five members (depending upon class size) to perform experiments and analyse the results during the eight weeks of intensive laboratory-based research. The laboratory session starts with "Discussion groups", made up with one member from each team. The aim of the Discussion Groups is for students to share their findings across teams and compare results from the previous sessions, reflecting the way in which scientists collaborate.

Instead of or in addition to classic demonstrators, we favour the use of paid "Peer Mentors", usually enthusiastic students from the previous year's cohort identified through a selection process. Peer Mentors facilitate the cross-team discussions and work closely with laboratory groups to bridge the gap between researcher and student. In this process, the Peer Mentors also benefit, as they are taught by the teaching team how to facilitate critical thinking rather than reveal 'answers'.

Experimental Outline

A clear outline of the experimental investigation to be performed is critical for the success of the "*Plant Detective*" activities. Your instructor will provide you with one of these. The actual order of the activities in this manual will be customized by your instructor to suit your course circumstances and will include dates and type of activities, necessary preparation, the teaching staff who will be present (Peer Mentors and instructors). Description of experimental procedures and age of the material are also important for the planning and execution of the experiments. Some of the experiments may differ in timing depending upon how your mutant plants develop. A second sowing, usually in week 2 (under supervision), is highly advisable because it gives you hands-on experience with plant growth and care and as well as providing backup material for extra experiments. For more information on Arabidopsis development see Appendix D. The experimental outline will help you gaining a broader outlook of the project.

Make the most of your research efforts!

(A statistician and former Plant Detective's perspective)

Before you embark on this Plant Detectives Project it is worth taking a few minutes to think about some key principles in experimental design and analysis. The project you are about to do has ample opportunity for independence and creative thinking. To make the most of your efforts it is important to give some thought to how you design your experiments and collect your data. Along the way you will also learn about how to analyse your data. The quality of your results, however, flows directly from the care that you have taken in experimental design and data collection.

Statistics is a philosophy as well as a quantitative tool. Modern statistical thinking in laboratory experiments is less than 100 years old. Much of the seminal work in developing the philosophy of modern statistics is credited to R.A. Fisher, also a leading figure in the field of genetics. Fisher was based in Rothamsted Experimental Station from 1919 to 1933, a research institute that continues to be a hotbed of agricultural research. It was during this time that he codified principles of experimental design and analysis. Unfortunately his writing is somewhat turgid and while his seminal work (*Statistical Principles for Research Workers* (1925)) is a classic, it is not read as widely as it should be.

There are some important basic principles in experimental design and analysis that are central to statistical thinking. These principles follow from the understanding that many uncontrolled factors influence the outcome of experiments. These factors include, for example, the effects of varying temperatures, humidity, greenhouse positioning, soil type and variation among individual plants. Following the basic principles can help researchers to design more efficient experiments and eliminate systematic biases that may produce misleading results.

1. **Controls** direct comparison against a known standard. A control may consist of absence of treatment, sham treatment or placebo. Control should be concurrent and tested under identical conditions to treatment. In this set of experiments, the controls are the wild type plants (or leaves or seeds thereof).

   Whenever you set up an experiment, ask yourself: How can I ensure that wild type and mutant seedlings (in pots or germination plates) are exposed to identical conditions.

2. **Blocking** means evaluating genotypes/treatments under homogeneous (low variability) experimental conditions. Treatment effects are measured within blocks and averaged across blocks. A block can be a set of units that share characteristics that may influence outcomes. Examples include a tray of seedlings, a technician responsible for processing a set of units, or units that are processed in a batch or run. Your data has 'structure' and your plants have a history, like a character in a novel has a back story — blocking can help you to keep a record of that history (e.g., where you place them, how you arrange your genotypes and treatments, who measures what) so that you have the option to account for it in your analyses. Utilising blocking can improve precision and lead to more efficient experiments.

   For example:
   a. When you begin your germination assay, determine whether you have a blocked design.
   b. When arranging your plants in trays, ask yourself: How can plants in this tray be arranged to avoid a systematic bias in outcomes?
   c. You are working in a group of other researchers — this is another source of bias. Discuss amongst yourselves how you can best avoid a systematic bias in outcomes?

3. **Replication** relates to the number of independent units within a treatment/genotype. The more independent units in the randomised experiment, the more information there is about the treatment, and the more precisely the treatment effect can be estimated. Repeating a measurement on a single unit can also add information about the treatment effect, but this is often not as informative as collecting the same number of measurements in independent units. For example, a leaf collected from a single plant that is split into three samples carries less information than leaf samples from three separate plants. Likewise, technical replicates in a spectrophotometer are not as informative as replicates from separate plants (also called biological replicates).

   It is important to distinguish these types of replicates in the statistical analysis. If not, one risks making incorrect inference. In the data analyses of this practical, you will learn how to analyse these data using Analysis of Variance (ANOVA).

   a. When collecting leaf samples for pigment assays, how will you sample leaves to ensure that you obtain three independent samples from each genotype?

   b. In the germination assay, what are the independent units?

4. **Randomisation** relates to a probabilistic process in which independent units are assigned to treatments. Randomisation helps to remove biases that may arise from uncontrolled factors, such as position in the tray or on the germination plate. In this practical, there are two situations where randomisation can be employed. Plants can be randomised to positions within a tray. In addition, when half of the plants of each genotype are chosen to undergo drought conditions, a randomisation algorithm rather than purposeful selection can be utilised. When randomisation is used, one can infer that any statistically significant difference between groups can be attributed to the genotype or treatment received.

   For example:

   a. What potential biases could arise if mutant seedlings were always on the left side of the tray and wild type seedlings on the right side?

   b. What randomisation method could be used to select which plants will receive drought/normal conditions?

5. **Blinding** masks the treatment assignment during assessments. Did you know that it is a well-observed fact that studies that are unblinded report larger treatment differences than studies that are assessment blinded? You can overcome this problem in your experiments.

   For example:

   a. What system can you and your colleagues use when collecting the phenotypic data in the seedling experiments? How can you ensure that the evaluator remains 'blind' to the genotype of the plant that is being measured?

   b. Similarly in the phenotypic assessments in the germination plate assays, how can the evaluator remain blinded to genotype?

Dr. Terry Neeman

The Australian National University Statistical Consulting Unit

(and former plant detective)

Activity 1: Observing the plant phenotype

1.1) Introduction and objectives

A common and simple way to assess the effect of gene mutations is to observe the resulting phenotype. The phenotype is the expression of the underlying genetic information and reflects the interaction between genes and the environment. Describing the phenotype of your wild type and mutant plants over the course of their development improves your ability to infer the biochemical pathways or physiological processes affected by the mutated gene.

Plant growth analysis has been defined as an explanatory, holistic and integrative approach to interpreting plant form and function (Hunt *et al*. 2002). The classical, *destructive method* involves tissue harvesting at regular intervals and estimation of different physical parameters, such as dry weight of plant parts (leaves, stems, roots and reproductive structures), leaf areas and volumes. These primary data are then used to calculate other parameters, such leaf area index, root/shoot ratio, and relative growth rates, in order to study plant morphology and function. 'Destructive methods' are widely used and are effective. They are resource (i.e., many plants are required) and time intensive, however, and will only be used towards the end of the Plant Detectives Project (in Activity 9, when relative water content is measured).

An alternative method to study plant growth and development is the use of *non-invasive* techniques. These are based on the direct measurement of leaf number, size and height, or on periodical image recording. Measurements are performed under controlled conditions and often involve subsequent software analyses. Image-based growth analysis is becoming a common approach as advances in robotic technology become broadly accessible. Image analysis has the advantages of requiring less starting material than the classic destructive method and also enabling a given individual to be monitored throughout its life cycle. Moreover, the photographic record can be analysed at any time in different ways so as to look for other morphological traits.

The main goals of Activity 1 are to:

1. follow the growth and development of wild type and mutant plants for several weeks. You need to familiarise yourself with the general morphology and organs of Arabidopsis (Fig. 1C) and the developmental growth stages described in Appendix C: Arabidopsis growth stages

2. use the LemnaTec Scanalyzer 3D (LemnaTec, www.lemnatec.com) to capture an image at one time point and analyse the major morphological features, such as leaf area, rosette circumference and leaf area coverage (Fig. 2). The image capture unit allows for image recording in reproducible conditions once the light and camera parameters are set up

3. analyse digital images. Digital images will be analysed based on object recognition and colour classification using the software, and morphological parameters will be compared.

The objective of Activity 1 is to compare the phenotype of the wild type versus mutant plants using traditional techniques (e.g., visual inspection, measurements of plant parts and leaf pigment observations) and image analyses.

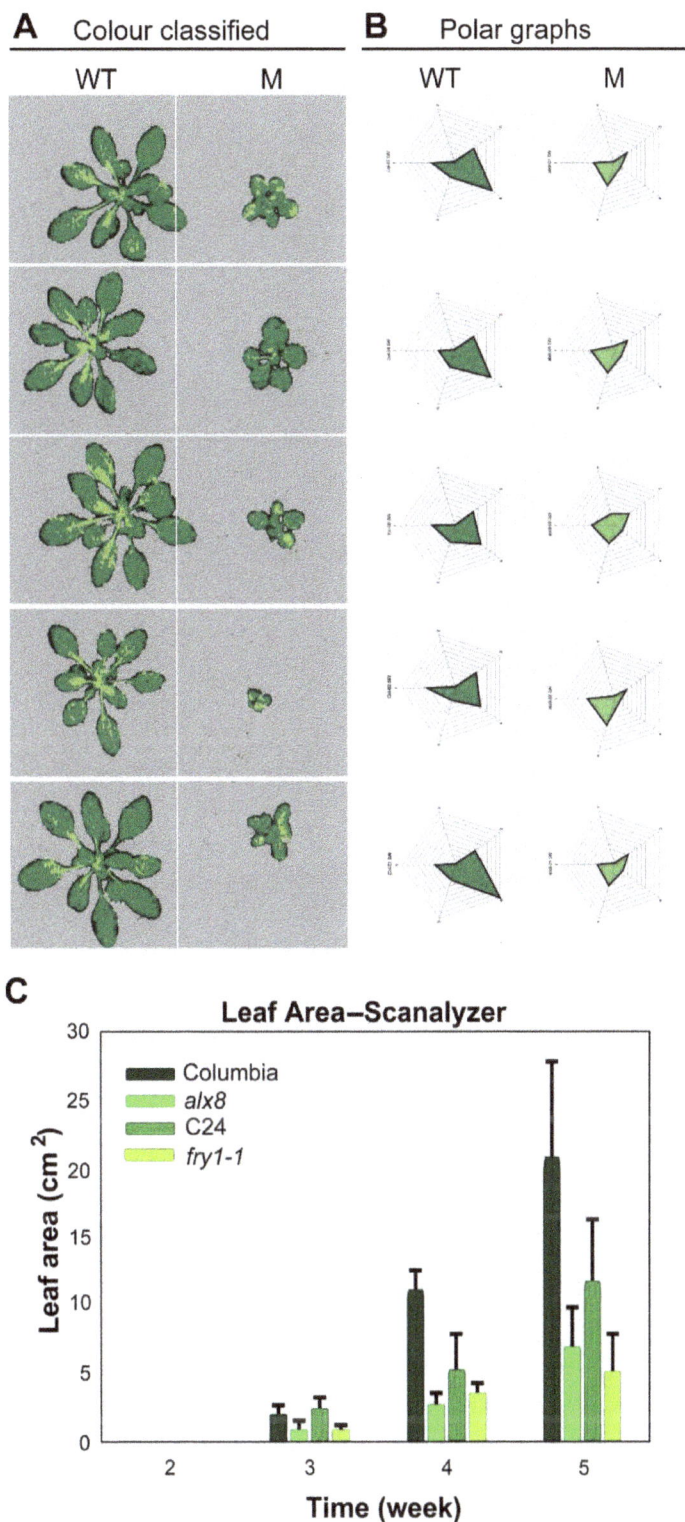

Figure 2. Non-invasive analyses of plant morphology
A) Images taken for five-week old wild type (WT) and mutant plants (M) and converted to false-colour using the LemnaTec system. B) Polar graphs representing five measured parameters of plant growth: relative growth rate (RGR), eccentricity, compactness, roundness and surface coverage. The data were normalised to the highest value in the series (axes scale is 0 to 1). C) Leaf area growth over time for four different genotypes; Columbia and C24 are wild types; alx8 and fry1–1 are mutant alleles (same gene) in the Columbia and C24 backgrounds, respectively.

IMPLEMENTING GOOD DESIGN PRINCIPLES IN ACTIVITY 1:

PLANT DETECTIVE TIP

1. MAINTAIN YOUR **DATA TRAIL**: MAKE SURE THAT YOU LABEL EACH PLANT AND RECORD WHAT TRAY IT SITS IN AND WHICH MEASUREMENTS COME FROM THAT PLANT.

2. **RANDOMISE** THE PLACEMENT OF WILD TYPE AND MUTANT PLANTS IN THE TRAYS (SEE PREFACE), BE MINDFUL OF BLOCKING.

3. **SOURCES OF VARIATION**: REMEMBER THAT YOU, AS THE OPERATOR, ARE A SOURCE OF VARIATION. DON'T LET ONE PERSON DO ALL WILD TYPE MEASURESMENTS AND ANOTHER DO ALL THE MUTANTS!

4. **BLINDING**: TRY TO ENSURE THAT THE PERSON DOING THE MEASUREMENT DOES NOT KNOW THE PLANT'S GENOTYPE.

5. **RECORD YOUR DATA** IN YOUR SPREADSHEETS, AND COLLECT PHOTOGRAPHIC EVIDENCE AS WELL (SEE BOX ON QUALITATIVE AND QUANTITATIVE DATA). MAKE SURE YOU HAVE YOUR SPREADSHEETS PREPARED IN ADVANCE AND HAVE ALL THE RELEVANT COLUMNS (SEE PREFACE AND APPENDIX B).

6. **REPLICATION**: MAKE SURE YOU MEASURE ENOUGH PLANTS TO GET A REASONABLE SAMPLE SIZE! YOU NEED AT LEAST THREE SAMPLES TO CALCULATE VARIABILITY!

1.2) Materials

The materials required for this activity are:

1. camera and tripod
2. cold room or fridge to stratify the seeds and coordinate germination
3. commercial seed-raising mix (mixes such as Debco work well); pasteurised soil is preferred to deter pest and pathogen growth
4. slow-release fertiliser (e.g., Osmocote Exact Mini at 3 grams (g) of fertiliser per litre (l) of soil; mix after the soil has been pasteurised)
5. filter paper
6. forceps
7. growth chamber
8. magnifying glass
9. mesh, non-slip PVC mesh (e.g., Matting Non Slip Magic Stop 90 centimetre (cm) Natural from Bunning's) — optional
10. plastic labels or coloured sticky tape to label individual pots
11. plastic lids or cling film
12. 48 pots per group (e.g., 63 millimetre (mm) Square Squat individual pots — Garden City Plastics Cat# P63SSQ)

13. ruler
14. seeds from wild type and mutant plants
15. table outlining growth and development (Appendix C: Arabidopsis growth stages) and Excel spreadsheets (see Appendix B: What do I do with my data?)
16. 31 x 44 cm plastic trays for storing your plants

1.3) Procedure

On the first day of the practical you will start from Step 4. Remember to always keep safe conduct in mind as you work (See Appendix A: General rules for safety and conduct).

Materials	Method

Soil, pots, fertiliser	1. Mix Osmocote and soil at ratio of 3 g of Osmocote per l of soil. IMPORTANT:
	   a. 4 l of soil are enough for a 24-pot tray
	   b. add the soil to the pot and press gently
	   c. a cement mixer can be used for larger volumes of soil (leftovers can be retained for Step 3).

Tray	2. Place 24 pots per tray. TIP: it is a good practice to line the tray with a thin PVC mat to improve drainage.
Water	3. Add 1 l of tap water to the tray and let it soak overnight. If you see the soil level has gone down overnight you can top it up with the soil left over from Step 1. The pots are now ready for sowing.
Labels	4. Number labels from 1 to 48. Note the labels are indicated as 'WT' (wild type) or 'M' (mutant). It is best to have consecutive numbering to refer to individual plants (See 'Labels', Appendix B: What do I do with my data?)
Seeds, filter paper, trays, soil, Osmocote, pots, labels	5. Sow seeds by placing them on a piece of filter paper and tapping them onto the soil. Be careful not to contaminate adjacent pots. Try to add no more than five–ten seeds per pot, and make sure they are not clustered in one spot. Sow seeds in half of the pots for each of the wild type and mutant lines. IMPORTANT: label each pot with a unique number ('identifier'). You can link this number to a spreadsheet indicating genotype and condition.
	6. Randomise the location of the pots in the tray — i.e., don't place all the mutants on one side and all the wild types on the other. Perhaps discuss with your group as to why we suggest this (make the most of your research efforts!).
Plastic lids or cling film, cold	7. Cover the trays with a plastic lid or cling film and stratify at 4 ºC in the dark for 48 hours. This will help synchronise seed germination.
	8. Move trays to growth chamber or glasshouse. Add ~800 millilitres (ml) of tap water to each tray every other day (check with your instructor as this may be done for you).
	9. Remove plastic lids on day ten or 12.
Forceps	10. Thin the plants to one–two per pot after the first two to four leaves are fully expanded (usually 14 days). IMPORTANT: don't discard the plants you remove: you can use the thinned plants to examine below-ground phenotype — see Step 13.
Camera, ruler	11. Observe the growth and development of the plants every week throughout the practical. Feel free to take photos or any other type of observation at any time. TIP: always place a ruler near the object when taking a photo to use as a scale bar.
	12. The Scanalyzer imaging schedule will depend on resources available and your goals. a. A single time point imaging of both wild type and mutant trays will allow you to compare morphological parameters.

Materials	Method
	b. Multiple time points imaging (at least four times) between weeks three and five will allow you to calculate relative growth rates in addition to morphological traits. Note that all plants will be imaged at the same time and that should give you enough statistical power to find significant differences.
	13. Analyses of roots. You can do this assessment on the seedlings you remove during thinning, as well as on the plants you harvest at the end. Periodic harvests along the way can also be informative, provided you have sufficient plants. To analyse the roots, VERY carefully remove them from the soil and gently wash under water. Arabidopsis roots are thin and fragile, so special care is required when cleaning them. Roots can be floated on a large Petri dish with water, scanned and analysed, as in Appendix E: Software tips.

Alternative procedures for morphological studies

1. Trace individual leaves on a piece of paper. Cut the paper. Compare the weight of all pieces of paper per plant to a weighted piece of paper of known area. You will be able to estimate total leaf area by comparing the weight of the 'paper' leaves to that of the standard piece of paper.
2. See Activity 9, Section 9.3.6 for more ideas about how to measure leaf area.

1.4) Expected outcomes

Record your observations about growth and development from *at least* five plants per genotype. Measure the same plants at each measurement date. Use your spreadsheet to identify which plant each measurement came from. The following lists some of the morphological parameters to observe and how to record them:

1. number of leaves. Note that the first leaves, or cotyledons, are not considered to be 'true' leaves
2. leaf shape, size and morphology; are leaves similar in form and shape?
3. pigmentation; is there any difference in the colour of plant parts when you compare the wild type and mutant?
4. rosette diameter; what plants have larger rosettes?
5. stem (bolt) height, as measured from the base to the tip of the stem
6. cauline leaves (number and length): the leaves produced in the stem, above the rosette
7. flower/silique number and size
8. the relative growth rate can be calculated RGR = $(\ln W2 - \ln W1)/(t2-t1)$, where W1 and W2 are plant dry weights or leaf area at times t1 and t2 (Hoffmann and Poorter 2002).
9. root traits, such as length, mass, colour and branching pattern
10. digital images obtained with the Scanalyzer system, see Appendix E: Software tips (Arvidsson *et al*. 2011), for some options for data analyses. An example of data presentation is given in Fig. 2.
11. graph and analyse your data, see Appendix B: What do I do with my data?

Note that the choice of parameters to measure is based on your preference and will depend on your observations. If you see something that suggests a difference between the wild type and mutant, take objective measurements of it. You are welcome to discuss with your instructors and peer mentors what parameters you would like to measure. Note that the plants will also be used for the other experimental assays, so keep your destructive assessments to a minimum.

QUALITATIVE VS QUANTITATIVE DATA

PLANT DETECTIVE TIP

1. QUANTITATIVE DATA ARE MEASURES THAT CAN BE COMPARED USING STATISTICAL ANALYSES

2. QUALITATIVE DATA INCLUDE PICTURES, DRAWINGS AND OBSERVATIONS

   A. THEY CANNOT BE COMPARED USING STATISTICAL TESTS, BUT CAN BE INVALUABLE RECORDS FOR YOUR MEMORY, FOR PRESENTATIONS, AND TO ILLUSTRATE THE DIFFERENCES THAT YOU TEST FOR USING THE QUANTITATIVE DATA

   B. FOR EXAMPLE, A TEST SHOWING STATISTICAL DIFFERENCES BETWEEN PIGMENT LEVELS CAN BE ILLUSTRATED WITH A PICTURE OF THE DIFFERENT THE COLOURATION OF THE LEAVES.

TIP: Quantifying qualitative traits

Often you will observe phenotypes that cannot easily be measured and quantified. For example, leaves might have slightly different colours or patches of coloured areas on them (Fig. 3A). There are different options to quantify such phenotypes:

1. If you know what the pigments might be, you could try and extract them with organic solvents, and quantify the amounts per leaf using high-performance liquid chromatography (HPLC), as explained in Activity 5.
2. Alternatively you may use qualitative assessment (Fig. 3B): Give each leaf a rank based on size and colour of the spot. Arrange leaves by rank, noting which one is wild type or mutant (it might be best to do this 'blind'; i.e., work with a colleague who removes the labels as he or she hands you the pot, or write on the back of each leaf if it is wild type or mutant). You can use the rank order to do a rank sum test, such as the Mann-Whitney U test, to determine significant differences between wild type and mutant.

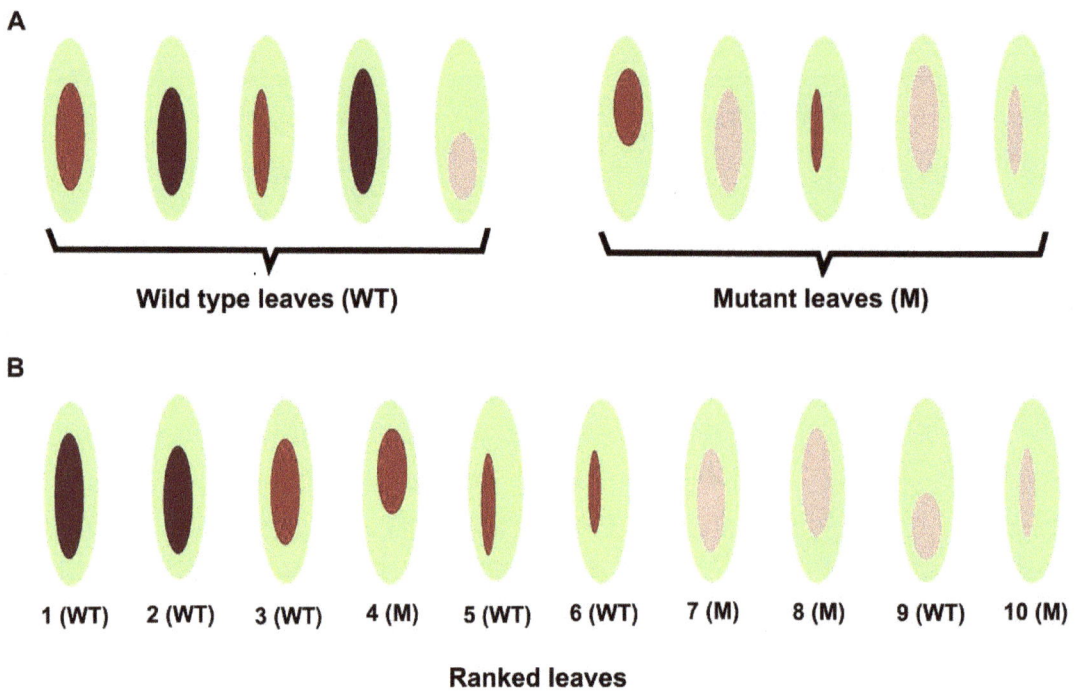

Figure 3. Quantifying qualitative traits.
A) Schematic representation of five leaves from wild type and mutant plants with different colours or patches of coloured areas on them. B) The same leaves arranged on a relative rank based on size and colour of the spot. You can use the rank order to do a rank sum test, like the Mann-Whitney U test, to determine significant differences between wild type and mutant.

Activity 2: Seed germination and root growth

2.1) Introduction and objectives

Seed germination is controlled by genetic and environmental factors. Some, such as Arabidopsis seeds, require humid conditions and germinate only in the light. Other seeds will only germinate in the dark or following a series of moisture or temperature changes. Seed germination is regulated by hormones, specifically abscisic acid (ABA), which is an inhibitor of germination, and gibberellic acid (GA), an inducer. Mutations in genes that regulate germination could affect either its onset or timing. Seed germination and early development phenotypes are useful during mutant screening.

Arabidopsis seeds exhibit a two-step germination, in which the seed coat (testa) ruptures first, followed by the rupture of the endosperm by the radicle (Muller *et al*. 2006). The tip of the root consists of rapidly dividing cells; these divisions cause the initial (or primary) root to extend the longer axis of the root. After a few days, thinner, 'lateral' roots emerge radially from the primary root (Fig. 4B) and epidermal cells of the primary root will differentiate into root hairs, thereby increasing total absorption surface.

Specific structures called statoliths, which are starch-accumulating amyloplasts, are present within specialised cells found in root tips (statocytes). Statoliths settle in the direction of gravity and are responsible for the 'gravitropic response' (Fig. 4D). This response senses the pull of gravity and directs root growth downward (Morita and Tasaka 2004).

You will monitor the germination, root growth and gravitropic response of seedlings grown on plates over the next two weeks. See more information about seed germination and root growth in the links provided in Appendix D: Databases and web resources.

The objectives of Activity 2 are to:

1. determine the germination rates of wild type and mutant seeds as the percentage of germinating seeds relative to the total number seeds plated
2. describe and quantify the gravitropic response of wild type and mutant seedlings
3. contrast patterns of root growth and morphology between wild type and mutant seedlings.

2.2) Materials

Fig. 4A shows the experimental set up for Activity 2. Arabidopsis seeds were sterilised and plated in transparent Petri dishes containing autoclaved agar medium supplemented with Murashige-Skoog medium (MS salts). Seeds were stratified (incubated in the dark at 4 ºC) to coordinate germination to occur four to six days before being transferred to the light. After germination, plates were placed vertically and root growth monitored. A detailed protocol for the preparation of the plates is provided in Appendix F: Seed sterilisation and plating or http://www.youtube.com/watch?v=MdnHvzON5ak&feature=youtube_gdata_player.

The materials required for this activity are:

1. growth chamber
2. laminar flow cabinet, sterile seeds, filter paper, pipette, tips

3. Micropore tape or parafilm
4. wild type and mutant seeds (~ 50 of each)
5. per group; four square Petri dishes with 20 millilitres of sterile 1.2% (w/v) agar and 0.5X Murashige-Skoog (MS) salts. Circular Petri dishes work, but can limit the number of seeds
6. ruler
7. sharpie marker
8. scanner or camera to record plates (optional)
9. styrofoam racks to hold plates vertically

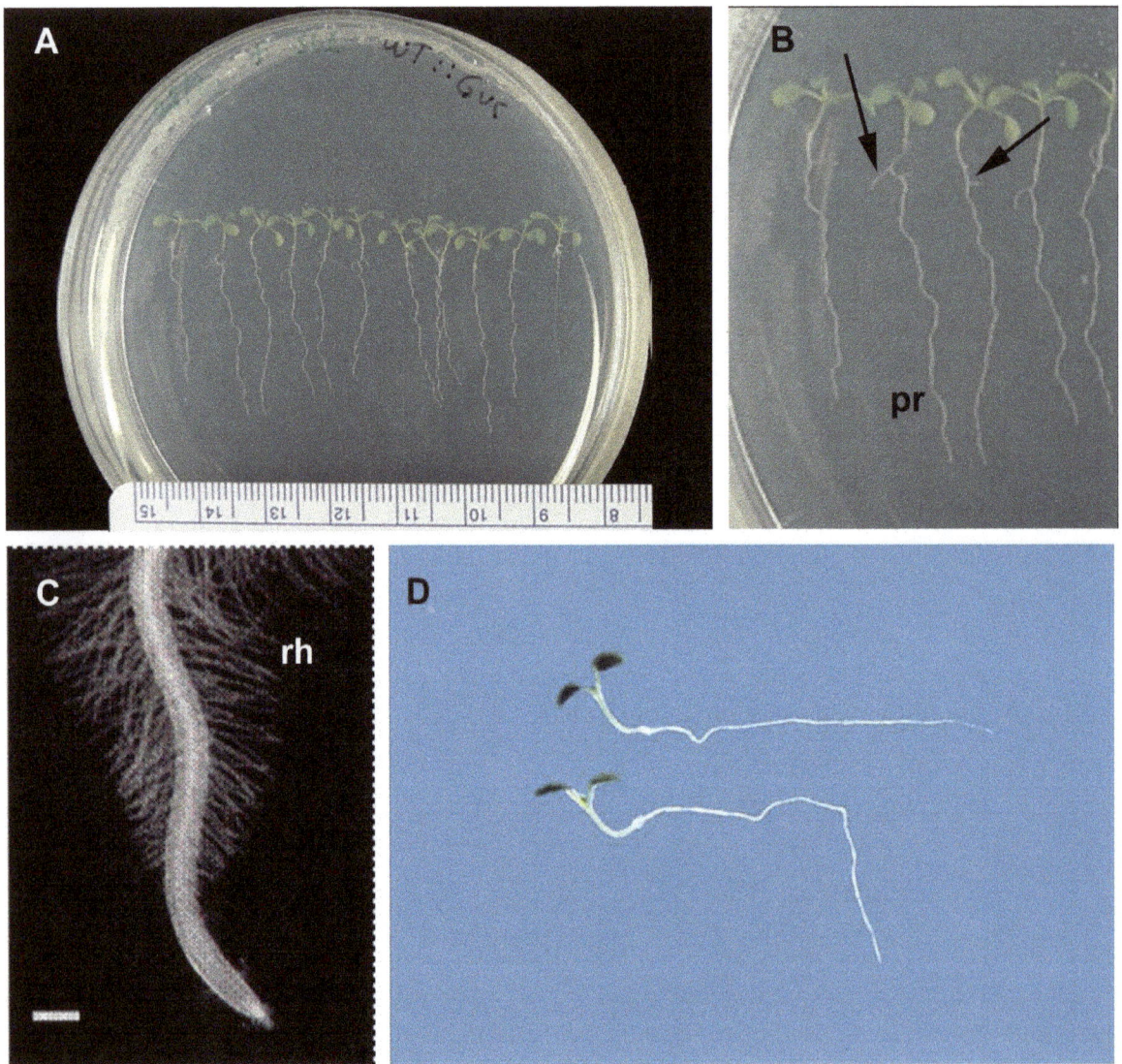

Figure 4. Investigation of seed germination and root growth
A) Arabidopsis seeds were sterilised and plated in autoclaved agar medium supplemented with medium (MS salts). After germination under light, plates were placed vertically and root growth monitored. Photo of 12-day-old seedlings. B) Seedlings with primary root (pr) and lateral roots (arrows). C) Root tip and root hairs (rh). D) Gravitropic response of a root after being placed horizontally. Image A courtesy of Peter Crisp (The Australian National University); Image C courtesy of Humboldt University; Image D courtesy of Whitehead Institute (MIT).

IMPLEMENTING GOOD DESIGN PRINCIPLES
IN ACTIVITY 2:

PLANT
DETECTIVE
TIP

1. MAINTAIN YOUR DATA TRAIL: KEEP TRACK OF PLATE, POSITION ON PLATE, AND GENOTYPE FOR EACH PLANT.

2. BLOCKING: KEEP IN MIND THAT YOUR PLATES ARE YOUR BLOCKS IN THIS EXPERIMENT (SEE PREFACE). ANALYSE WITHIN THE PLATE BY USING PLATE MEANS, THE SEEDLINGS ARE TECHNICAL NOT BIOLOGICAL REPLICATES.

3. RANDOMIZATION: TO THE EXTENT POSSIBLE THE GENOTYPES SHOULD BE RANDOMISED BETWEEN SIDES OF THE PLATES (OR PERHAPS THE DIRECTION THE PLATE IS FACING IN THE RACK).

2.3) Procedure

Note that steps 1 to 5 will be performed for you in advance. You need to start from Step 6.

Materials	Method
MS agar plates; sharpie	1. Label four to six plates with 'date, group name (A–E), plate (1–4).'
Ruler, sharpie	2. Trace one line with the ruler and sharpie at roughly 2/3 of the plate side length.
Laminar flow, sterile seeds, filter paper, pipette, tips	3. Using a long glass Pasteur pipette or regular-tip pipette, plate six sterile wild type seeds on one side and another six seeds of the mutant on the other side. If possible, randomly assign side of plate. Label with thin sharpie accordingly.
Micropore tape	4. Seal the plates with strips of Micropore tape or parafilm.
	5. Stack the plates on top of each other, wrap with aluminium foil and incubate them in the dark at 4 ºC for five days to coordinate germination.
Styrofoam rack, growth chamber	6. Place the plates vertically on the rack provided and transfer them to the growth chamber with a light intensity of ~120–150 micromole (µmol) m^{-2} s^{-1}. Plates can be incubated under continuous light or for a 16-hour photoperiod. TIP: make sure the light source is from the top and directly perpendicular to the plates during germination and early growth (before Step 10).
	7. Record total number of seeds, and how many germinate after three days in the light.
	8. Monitor root growth and development for two weeks.
Ruler	9. Follow root growth rate by marking the position of the tip (and only the tip) of the main root on the plate surface with a Sharpie at periodical intervals (two–four) days. Also note root direction (angle) and secondary/lateral root density.

Materials	Method
	10. Flip/rotate orientation of two plates 90 ºC after seven days, so that the growing seedling roots are now oriented horizontally. Over the next several days observe root orientation response. Measure root lengths before and after flipping (see Step 11).
	11. Images of plates can be recorded with a scanner or camera. TIP: a scanner is usually more effective. Either way, place a black background behind the plate to increase the contrast of the white roots. Include a ruler as a scale. Note that the plastic lid will interfere with the camera flash or focus, to avoid this, plates may be imaged from the bottom face. Check image quality before proceeding with further recording.

2.4) Expected outcomes

1. Record the number of germinated seeds per genotype to calculate the germination rate. We will score germination as the endosperm rupture evidenced by the emergence of the radicle through the testa after three, four and five days. If you record when each seed germinates you can also determine mean time to germination for each genotype in addition to total germination percentage relative to the initial number of seeds sown. You should count each plate separately and record data. You will then take the mean of the values across your plates so that you can do the appropriate statistics.

2. Monitor the root length over time. This will allow calculating root growth rate as the change in length over time (millimetres/day). The easiest way is to measure the distance between the line on the plate and the tip of the primary root. More accurate and quantitative root analyses can be achieved by scanning or photographing the plates and using the free EZ-Rhizo software package (Armengaud et al. 2009) or ImageJ, as explained in Appendix E: Software tips. If you choose to do this, use your phone or a camera to take pictures of the plates — make sure to include a ruler on the side of each photo for scale.

3. Describe and compare root morphology of the two genotypes (i.e., primary root, growth direction, and number of secondary (lateral) roots).

4. Record any other observations that you consider relevant.

5. Determine if there is a statistically significant difference in the above parameters between the wild type and the mutant (see data analysis Appendix B: What do I do with my data?).

6. Remember to record your data in your spreadsheet and to record the plate number for each measurement you take.

Activity 3: Extraction and quantification of photosynthetic pigments

3.1) Introduction and objectives

Although most plant leaves appear to be green, several different colour pigments are usually present in the chloroplasts of green leaves (Fig. 5). Colours are a result of light absorption by pigments at specific wavelengths. These pigments are not simply decoration, but play critical roles in photochemistry — photosynthesis and photoprotective mechanisms. Mutations that affect pigment composition of leaves can have substantial functional impact, and can be used to explore the genetic control of pigment biochemical pathways and physiological processes.

The chlorophylls a (Chl*a*) and b (Chl*b*) provide the green colour and absorb the light energy that is needed for photosynthesis. Chl*a* has a methyl group in the position where Chl*b* has a formyl group, which gives Chl*a* slightly more affinity for non-polar solvents relative to Chl*b*. Pheophytin plays a vital role in electron transport, but also may occur as an acid-induced breakdown product of chlorophyll. This other form of chlorophyll lacks the central Mg^{2+} that Chl*a* and Chl*b* have; this results in a higher affinity for non-polar, hydrophobic solvents.

Closely associated with the chlorophylls in the chloroplast is another group of pigments, the carotenoids. They are yellow to red in colour and likely play a role in the gathering of light energy for photosynthesis, as well as helping to protect the chlorophylls against photo oxidation. The carotenoids are divided into two groups: the carotenes, a pure hydrocarbon group; and the xanthophylls, which are characterised by two additional oxygen atoms. The additional oxygen, which is present as hydroxyl, groups at the ends of the molecule, making the xanthophylls more polar. In other words, the xanthophylls have less affinity for non-polar solvents than the carotenes.

Differences in the chemical structure of pigments not only determine their absorbance spectrum and, thus, their function, but they also change the pigment affinity for different solvents. This difference in solubility allows the separation of pigments by chromatographic techniques and the specific absorbance spectrum can be utilised to identify individual pigments (Fig. 6C).

The main objectives of Activity 3 are to:

1. isolate total pigments from both wild type and mutant plants grown in the light using a simple organic solvent extraction procedure
2. quantify pigments based on their spectral properties.

3.2) Materials

Numbers in parenthesis indicate number of specific items per group.

1. 1.5 millilitre (ml) Eppendorf tubes and rack
2. 80% (v/v) acetone in capped bottle
3. block heater (optional)
4. gloves
5. microcentrifuge (one per class or one per group) (swing out bucket is best, but not indispensable)
6. micropipettes of 200–1000 microlitres (μl) and tips
7. safety goggles (one per person)
8. vortex (one per class or one per group if available)

9. plate reader BIO-TEK uQuant and 96 well plate or spectrophotometer and cuvette
10. waste container for acetone in the chemical hood
11. liquid nitrogen (−197 °C!!!)
12. plastic pistils for grinding tissue or ball bearings and a TissueLyser

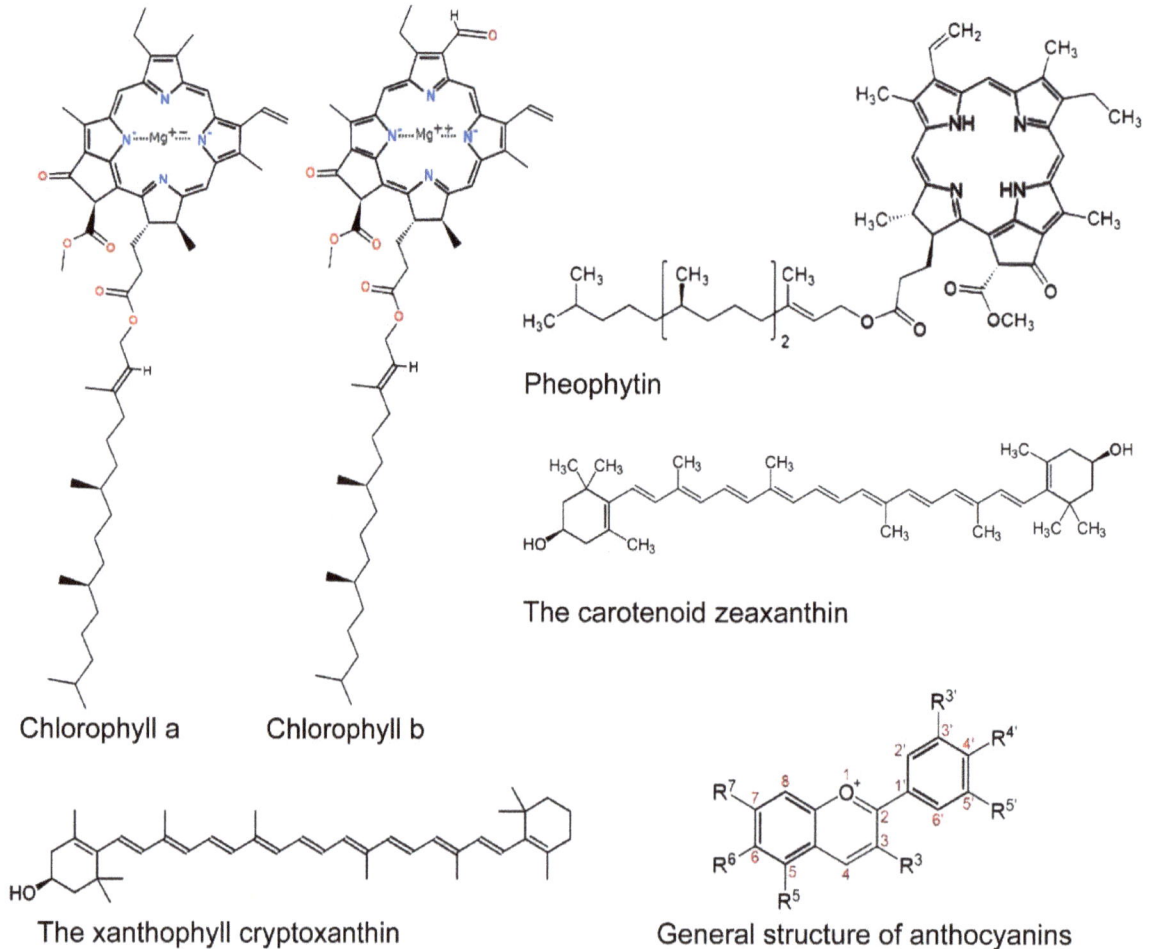

Chlorophyll a   Chlorophyll b

Pheophytin

The carotenoid zeaxanthin

The xanthophyll cryptoxanthin

General structure of anthocyanins

Figure 5. Structure of plant pigments

3.3) Procedures

3.3.1) Pigment extraction for spectrophotometric analyses

Plant pigments can be extracted from plant tissues by organic solvents. The tissue is first frozen in liquid nitrogen and ground, the ground tissue is then treated with organic solvents to extract the plant pigments. Using the protocol below you will extract Chl*a* and Chl*b* and quantify their content in your leaf tissues. Depending on who performs the harvest and the type of grinding method, you may be asked to continue from Step 3 or Step 4.

IMPLEMENTING GOOD DESIGN PRINCIPLES IN ACTIVITY 3-6:

PLANT DETECTIVE TIP

1. **REPLICATION** AND SAMPLING BIAS: CAREFULLY CONSIDER YOUR SELECTION OF LEAF SAMPLES. BULKING ACROSS LEAVES MAY BE NECESSARY, BUT ENSURE THAT YOU HAVE A RANDOM SELECTION OF LEAVES SO YOU ARE CERTAIN THAT EACH SAMPLE IS AN INDEPENDENT BIOLOGICAL REPLICATE. HOW CAN YOU ACHIEVE THAT?

2. **REPLICATION:** SOME OF THE EXPERIMENTS IN THIS ACTIVITY WILL HAVE A SAMPLE OF ONE FOR LOGISTICAL REASONS (N=1). REMEMBER THAT, WHILE NOT MUCH CAN BE EXTRAPOLATED FROM A SAMPLE OF ONE, THESE RESULTS CAN BE USED TO SUPPORT OTHER DATA.

3. **DATA TRAIL:** RECORD FROM WHICH PLANTS YOU TOOK EACH OF YOUR SAMPLES.

4. **BLINDING:** BE AWARE OF EXPERIMENTER BIAS WHEN TAKING MEASUREMENTS.

IMPORTANT

YOU WILL BE PROVIDED WITH CHEMICAL HAZARD INFORMATION. PLEASE READ THE RELEVANT SHEETS BEFORE STARTING THE EXPERIMENT. SEE APPENDIX G.
ALWAYS WEAR SAFETY GOGGLES AND GLOVES WHEN WORKING WITH LIQUID NITROGEN AND ORGANIC SOLVENTS THROUGHOUT THE ACTIVITIES IN THIS MANUAL.

Materials	Method
Eppendorf tubes	1. Harvest approximately 30 milligrams (mg) of leaf tissue from FOUR wild type and mutant plants and place each into a 1.5 ml Eppendorff tube. If leaves are small you may need to bulk samples. Keep design principles in mind. Make sure tubes are properly labelled beforehand and contain a bearing ball for the grinding if using the TissueLyser. Alternatively, you can grind tissue using plastic pistils (Step 3). See labelling tips in Appendix B.
Goggles, gloves, liquid nitrogen	2. Freeze the material immediately by dipping the sample in liquid nitrogen. IMPORTANT: wear goggles and gloves to avoid burning your skin. Liquid nitrogen boils at −196 °C (very cold!) and causes rapid freezing when in contact with living tissue. TIP: it is best to use a floating styrofoam rack to keep samples upright and avoid liquid nitrogen leaking into the tube.
Pistil for grinding	3. Grind the frozen tissue, taking care not to let the samples thaw. If thaw starts, dip the tubes in the liquid nitrogen again.

Materials	Method
Acetone, pipettes, tips, vortex, gloves	4. Add 1000 µl of 80% (v/v) acetone to the sample. Wear goggles at all times while working with solvents!
Vortex	5. Vortex for ten seconds. TIP: if it is necessary to increase extraction efficiency, incubate under light (if available) or at 37 ºC to 40 ºC for ten minutes, vortexing periodically.
Microcentrifuge	6. Centrifuge the sample for two minutes at 7000 g in a microcentrifuge.
	7. Carefully, transfer as much supernatant as possible to a new tubes. Be careful not to remove pellet. Usually, 700–900 µl can be recovered depending on how packed the pellet is at the bottom. Remember to label the new tubes accordingly!
	8. For one of the four tubes per genotype, transfer 500 µl to another tube labelled 'group#–WT' or 'group#–Mutant'. These two samples per group will be used in Activity 4. If they are to be used at another time they can be stored at −20 ºC.

3.3.2) Spectrophometric quantification

Dissolved pigments absorb light of specific wavelengths directly proportional to their concentration. This relationship is expressed quantitatively by the Beer-Lambert law:

$$Abs_\lambda = \varepsilon \times l \times C$$

where Abs_λ is the absorbance at wavelength λ, l is the light path length (centimetre (cm)), ε the millimolar extinction coefficient, or a molecule–specific constant in a given solvent (L g^{-1} cm^{-1}), and C is the concentration (g L^{-1}). We can calculate the concentration of any absorbing molecule if we know its absorbance at specific λ, l and ε. In this case, we need to use the ε for 80% (v/v) acetone.

Materials	Method
96 well plate	1. Add 200 µl of 80% (v/v) acetone in two wells as blanks.
96 well plate	2. Transfer 200 µl aliquots of your samples into two or three wells.
Plate reader	3. Select the wavelength of 664 nanometres (nm).
	4. Read the absorbance of your sample at 664 nm and record the number.
	5. Repeat reading for the wavelength of 647 nm.
	6. Recalculate the absorbance of your samples by subtracting that of the blank.
	7. Calculate chlorophyll concentration as: $[Chla](\mu g/ml) = 12.25 \times Abs_{664nm} - 2.55 \times Abs_{647nm}$ $[Chlb](\mu g/ml) = 20.31 \times Abs_{647nm} - 4.91 \times Abs_{664nm}$ Note that the ε values are those for chlorophyll in aqueous solution of 80% (v/v) acetone.
Waste container, chemical hood	8. Discard the extract in the proper waste container in the hood.

Alternative procedure for pigment characterisation using a regular spectrophotometer

The use of a plate reader gives more flexibility and higher replication power as many samples can be read and a whole spectrum (i.e., absorbance readings at different wavelengths) can be obtained in a matter of minutes. If a plate reader is not available, however, the same measurements can be performed with a regular single cuvette spectrophotometer. Make sure to run at least three replicate readings at 647 nm and 664 nm for each of the three biological replicates per genotype for the quantification of chlorophylls; the same equations can be used provided the solvent is 80% (v/v) acetone.

3.4) Expected outcomes

1. calculate the amount of Chl*a* and Chl*b* in a microgram (µg) of chlorophyll per milligram (mg) of fresh weight
2. calculate the Chl*a/b* ratio for all the samples. Graph and analyse the data and indicate if there is a significant difference between genotypes (See Appendix B).

Activity 4: Qualitative analyses of photosynthetic pigments by thin-layer chromatography (TLC)

4.1) Introduction and objectives

Chromatography is an analytical technique that allows for the separation and identification of a wide range of compounds. The separation of the components in a mixture is a function of their different affinities for a stationary phase, such as a solid or a liquid, and their differential affinity for a moving phase, such as a liquid or gas. When the stationary phase is solid and the moving phase is liquid, the separation of compounds is governed by their tendency to associate with the mobile (usually hydrophobic) liquid phase or to adsorb onto the solid (usually hydrophilic) surface. The solid phase might be paper, starch or silica gel. If the solid is applied in a thin layer to a supporting glass or plastic plate, the method is called thin-layer chromatography (TLC). In this protocol you will use pre-prepared glass TLC plates coated with silica gel. Visit http://prometheuswiki.publish.csiro.au/tiki-index.php?page=Thin+Layer+Chromatography to view the original procedure.

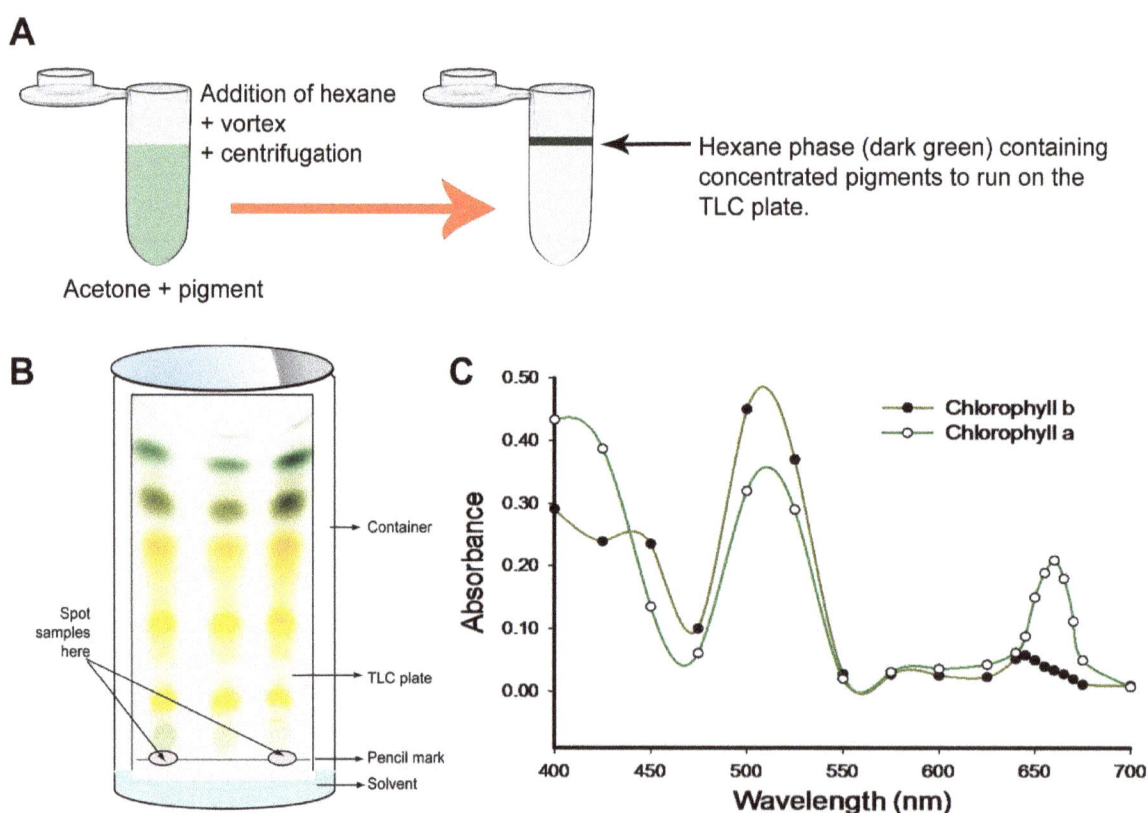

Figure 6. Pigment extraction and characterisation by TLC and spectrophotometry
A) Concentrating pigments prior to TLC. Pigments dissolved in the acetone extract can be concentrated by lowering the acetone concentration to less than 70% (by adding water) to facilitate the partitioning of the pigments to the hexane phase. After mixing and spinning the mixture, pigments will accumulate in the more hydrophobic, less dense organic solvent hexane (dark top phase). B) Schematic representation of a TLC run. Note that samples must be loaded on the pencil line. Load as much as possible to obtain darker, more concentrated spots. Pigments (yellow and green dots) will be resolved as the solvent migrates from bottom to top. C) Absorption spectrum of chlorophyll a (Chl*a*) and chlorophyll b (Chl*b*). Pigments were removed from the silica, extracted with acetone and their absorbance measured at regular intervals in the range 400 to 700 nanometres (nm). Note the distinctive Chl*a* peak at around 660 nm.

In TLC, the mixture to be separated is first applied as a spot or a line to the solid phase, and then the mobile solvent is allowed to pass through the applied compounds along the immobile phase (Fig. 6B). The compounds will dissolve in and move with the solvent. The distance travelled by a particular compound will depend on its affinity for the hydrophobic (mobile) phase versus its affinity for the hydrophilic (solid) phase, thus assisting with the identification of the compound. The ratio of the distance travelled by a compound to that of the solvent front is known as the Retardation factor (Rf) value. Unknown compounds may be identified by comparing their Rfs to the Rfs of known standards.

The different chemical structure of the plant pigment not only confers different absorption properties, but also different affinities for mobile and stationary phases that allows for their separation. Following the protocol below you will separate, identify and compare the pigment composition of your wild type and mutant plants.

The main objectives of Activity 4 are to:

1. separate individual plant photosynthetic pigments from both wild type and mutant plants
2. compare pigment composition between genotypes
3. measure the absorbance spectra of pigments to identify their nature.

IMPLEMENTING GOOD DESIGN PRINCIPLES IN ACTIVITY 3-6:

PLANT DETECTIVE TIP

1. REPLICATION AND SAMPLING BIAS: CAREFULLY CONSIDER YOUR SELECTION OF LEAF SAMPLES. BULKING ACROSS LEAVES MAY BE NECESSARY, BUT ENSURE THAT YOU HAVE A RANDOM SELECTION OF LEAVES SO YOU ARE CERTAIN THAT EACH SAMPLE IS AN INDEPENDENT BIOLOGICAL REPLICATE. HOW CAN YOU ACHIEVE THAT?

2. REPLICATION: SOME OF THE EXPERIMENTS IN THIS ACTIVITY WILL HAVE A SAMPLE OF ONE FOR LOGISTICAL REASONS (N=1). REMEMBER THAT, WHILE NOT MUCH CAN BE EXTRAPOLATED FROM A SAMPLE OF ONE, THESE RESULTS CAN BE USED TO SUPPORT OTHER DATA.

3. DATA TRAIL: RECORD FROM WHICH PLANTS YOU TOOK EACH OF YOUR SAMPLES.

4. BLINDING: BE AWARE OF EXPERIMENTER BIAS WHEN TAKING MEASUREMENTS.

4.2) Materials

1. heating block (or desk lamp)
2. capped bottle. This can be a ~250 millilitre beaker covered with a glass Petri dish. A microscope slide staining dish works best if TLC plates are of the proper size
3. forceps
4. gloves
5. goggles (one per person)
6. chemical hood
7. micropipettes of 5–50 microlitres (µl), or glass capillaries; check with demonstrator if you have questions about how to use these
8. pencil
9. ruler
10. scanner or camera
11. TLC plates (three) (i.e., TLC Silica gel 60 F_{254}, Merck Cat#115341)
12. waste container
13. razor blades
14. weighing paper
15. spectrophotometer or micro plate reader (i.e., BIO-TEK uQuant)
16. hexane: acetone: chloroform (2:1:1)
17. hexane

Warning:
HEXANE is highly toxic and should be handled in the fume hood.

4.3) Procedure

4.3.1) Pigment concentration

In order to load sufficient pigment sample in the TLC plate, it is necessary to concentrate the pigments that were extracted in Activity 3 in an organic phase (hexane).

IMPORTANT YOU WILL BE PROVIDED WITH CHEMICAL HAZARD INFORMATION. PLEASE READ THE RELEVANT SHEETS BEFORE STARTING THE EXPERIMENT. SEE APPENDIX G. ALWAYS WEAR SAFETY GOGGLES AND GLOVES WHEN WORKING WITH LIQUID NITROGEN AND ORGANIC SOLVENTS THROUGHOUT THE ACTIVITIES IN THIS MANUAL.

Materials	Method
Hexane, vortex, pipettes, tips	1. Add 80 µl of hexane and 100 µl of water to each of the **two samples** (one wild type, one mutant) that you aliquotted in Step 8 of Activity 3 in the tubes containing 500 µl of supernatant.[2]
Vortex	2. Vortex for ten seconds.
Microcentrifuge, Eppendorf tubes	3. Centrifuge four tubes for two minutes at 7000 g in a microcentrifuge to separate organic and aqueous phases.
Pipettes, tips, Eppendorf tubes	4. Transfer the upper phases (hexane) containing the concentrated pigments to new, LABELLED tubes (Fig. 6A). Keep the tube closed at all times as the hexane phase evaporates rapidly. These samples will be loaded onto the TLC plates next.

4.3.2) Thin-layer chromatography (TLC)

Materials	Method
Silica plate, pencil, ruler, gloves	1. **Wearing gloves**, mark the edge of each pre-prepared TLC plate with a pencil about 1 centimetre (cm) above the bottom, trying not to touch the silica gel surface. TIP: It is critical that the silica plates are dry before loading the sample to avoid smudges leading to poor runs.
Micropipette or capillary tube	2. With a micropipette or capillary tube spot the wild type sample on the left of the plate and the mutant one on the right of the plate. Try to make the spots evenly spaced on the pencil line.
Heating block, or lamp	3. Place the plate on the heating block to evaporate residual hexane. TIP: the sample can be also be dried by placing the slide underneath a desk lamp.
	4. Continue to load more sample on to the plate by repeatedly touching the silica with the micropipette or capillary containing 5–10 µl of the pigment. Let the spot dry for ten to 20 seconds and apply again. Be careful not to damage the silica gel layer.
	5. Load the sample at least **five** times, but be careful not to overload as the spots may overlap, resulting in cross contamination. Check with your demonstrator. IMPORTANT: **record** the number of loads so you know the amount of loaded sample.
Forceps, gloves, chemical hood, beaker covered with glass Petri dish, or microscope slide staining dish; hexane: acetone: chloroform (2:1:1)	6. Carefully, place the plate into the equilibrated chamber containing enough hexane: acetone: chloroform (2:1:1) mixture to cover the bottom and replace the lid. TIP: a beaker covered with a glass Petri dish, or microscope slide staining dish can be used for this step. IMPORTANT: ensure the plate is placed evenly to ensure the solvent comes into contact with the entire bottom of the TLC plate at once, otherwise the samples will not run evenly. Only the bottom of the TLC plate should be in contact with the solvent, NOT the sample. The solvent will be absorbed by the silica gel and will start to move up the plate through the spotted samples. When it reaches the pigment spot, the pigments will dissolve in the solvent and move with it along the silica gel.
Pencil	7. Once a clear separation of the pigments is achieved (check with demonstrator), remove the plate before the solvent front runs off at the end of the plate and immediately mark the solvent front with a pencil. The solvent will evaporate rapidly, so act **quickly**. Alternatively, you can use the end of the plate as your reference.

---

[2] The final concentration of acetone must be lower than 70% to facilitate the partitioning of the pigments to the hexane phase.

Materials	Method
Block heater or lamp, scanner or camera	8. Allow the plate to air dry on a block heater or under a lamp. At this point, you can scan or photograph the plate.
Waste container	9. When you are finished, pour the remaining solvent mixture into the waste container provided and leave the bottles in the chemical hood to dry.

4.3.3) Pigment identification

Although the pigments can be recognised by their colours, a more accurate way to characterise them is by measuring an absorption spectrum. In this section, you will remove the pigment bands for Chl$a$ and Chl$b$ (or any other pigment of your choice) from the TLC plate and measure their absorbance at different wavelengths. The amount of light absorption by a substance can be graphed as a function of wavelength; this graph is called an absorption spectrum (Fig. 6C). Absorption spectra are unique for individual pigments and should allow you to identify that molecule by comparing with published spectra.

Materials	Method
Razor blades, weighing paper	1. Remove the pigment bands from the silica plate you want to analyse by scraping the silica gel from the plate with a razor blade onto a square of weighing paper or aluminium foil.
Eppendorf tubes	2. Combine all of the scrapings for a given band together in one labelled Eppendorf microcentrifuge tube. Each tube will then contain all of a particular pigment present in the total loaded volume of hexane/pigment extract. (Remember that the total volume of the hexane/pigment phase was 80 µl; it is expected that ALL the pigments were partitioned from the acetone phase to the hexane one.)
Pipette and tips, 80% acetone, vortex	3. Extract the pigments from the silica gel by adding 500 µl of 80% (v/v) acetone, close the tube and mix well by vortexing.
Block heater	4. Incubate the silica/acetone mix under a light or in a heated bath for ten minutes, shaking periodically (notice the colour of the silica gel. Is the silica gel white or has some pigment been retained?).
Microcentrifuge	5. Sediment the mixtures for two minutes at 7000 g in a microcentrifuge to remove the silica gel.
Pipette and tips, 80% acetone, 96 well plate	6. Transfer 200 µl of 80% acetone into the wells as blanks.
Eppendorf tubes, pipette and tips	7. After pelleting the silica, call your supervisor before proceeding. Recover the cleared supernatant with a pipette and transfer to a properly labelled microcentrifuge tube. Transfer 200 µl of your samples into the wells as duplicates for each sample.
Plate reader	8. Transfer the plate into the plate reader. Read the blank and sample at 400 nm and every 20 nm until 800 nm.
	9. Calculate the 'Final' absorbance by subtracting the blank absorbance from the sample absorbance. Use the 'Final' absorbance and graph 'Absorbance' vs 'Wavelength' (nm) (Fig. 6C).

4.4) Expected outcomes

1. Characterise the pigment profile for both wild type and mutant plants. For this, use visual examination and also measure the Rf for each pigment spot. Rf is the ratio between the total distance travelled by the pigment and the total distance travelled by the mobile phase (solvent front). The starting point for each measurement is the 1 cm mark.

2. Identify each of the pigments using your knowledge of pigment structure and colour. The pigments visible on the chromatograms are Chl*a* and Chl*b*, carotene, xanthophylls (possibly more than one), and pheophytin. Based on time and your interest, you could examine all pigments or those that differ between wild type and mutant. Once you have identified the pigments, you can scan or photograph your plate and label each band.

3. Plot the absorption spectrum for the pigment of your choice.

4. OPTIONAL: compare the pigment profile among different genotypes and conditions (i.e., light-versus etiolated-grown tissue, if available).

Activity 5: Qualitative analyses of photosynthetic pigments by high-performance liquid chromatography (HPLC) — optional

5.1) Introduction and objectives

High-performance liquid chromatography (HPLC) is a technique used to separate, identify and quantify complex mixtures of compounds based on their solubility in a solvent compared to a stationary phase (Fig. 7). The principle is similar to thin-layer chromatography, but instead of running on the surface of a stationary plate, the compounds are separated on a column packed with a stationary phase, and solvent is pumped through the column at high pressure. HPLC provides better resolution of complex mixes than TLC, it is more quantitative, it can be coupled to an absorbance and fluorescence spectrophotometer and it produces reproducible results.

In our case, the total sample extract from wild type and mutant plants will be run through a narrow column at high pressure. The pigments are separated via hydrophobic interactions with the column matrix and detected by measuring absorbance. Their retention time, or the time a specific pigment takes to reach the detector, is used to identify each pigment.

The main objectives of Activity 5 are to:

1. isolate total pigments from both wild type and mutant plants
2. resolve different pigments by using HPLC
3. identify pigments by looking at the HPLC chromatogram and assess differences of the pigment composition between wild type and mutant samples.

IMPLEMENTING GOOD DESIGN PRINCIPLES IN ACTIVITY 3-6:

PLANT DETECTIVE TIP

1. **REPLICATION AND SAMPLING BIAS:** CAREFULLY CONSIDER YOUR SELECTION OF LEAF SAMPLES. BULKING ACROSS LEAVES MAY BE NECESSARY, BUT ENSURE THAT YOU HAVE A RANDOM SELECTION OF LEAVES SO YOU ARE CERTAIN THAT EACH SAMPLE IS AN INDEPENDENT BIOLOGICAL REPLICATE. HOW CAN YOU ACHIEVE THAT?

2. **REPLICATION:** SOME OF THE EXPERIMENTS IN THIS ACTIVITY WILL HAVE A SAMPLE OF ONE FOR LOGISTICAL REASONS (N=1). REMEMBER THAT, WHILE NOT MUCH CAN BE EXTRAPOLATED FROM A SAMPLE OF ONE, THESE RESULTS CAN BE USED TO SUPPORT OTHER DATA.

3. **DATA TRAIL:** RECORD FROM WHICH PLANTS YOU TOOK EACH OF YOUR SAMPLES.

4. **BLINDING:** BE AWARE OF EXPERIMENTER BIAS WHEN TAKING MEASUREMENTS.

A

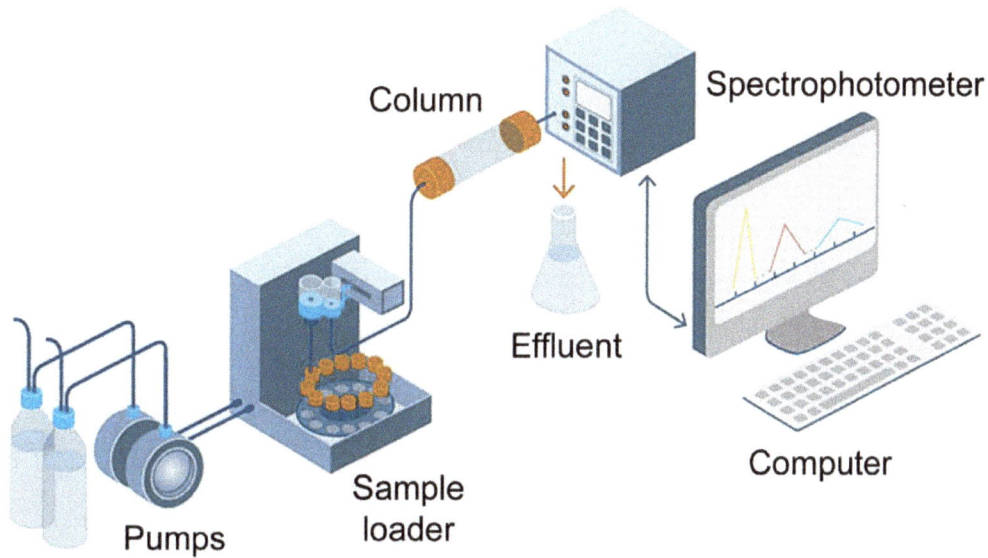

B

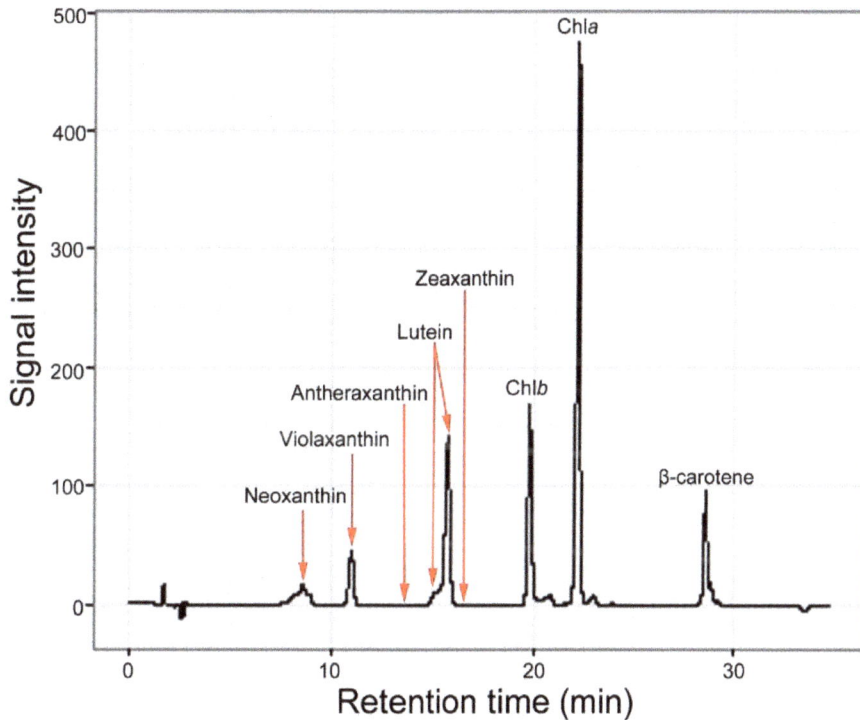

Figure 7. Analyses of plant pigments using HPLC

A) Schematic diagram of setup for HPLC. Two solvents (usually one polar and one slightly less polar) are mixed in a pump at certain ratios changing over time. The solvent mixture is pumped onto a column packed with stationary phase. The mixture to be separated is introduced onto the column and the solvents will separate the compounds, leading to distinct retention times (the time it takes for each compound to be eluted from the column). B) Each compound flows through an absorbance or fluorescence spectrophotometer and produces a typical absorbance or fluorescence spectra that can be used to identify each compound compared to a known standard. Each peak of this chromatogram from an Arabidopsis leaf sample represents a different pigment. The retention time of each peak is specific for individual pigments and the area under it is proportional to its abundance. Fig. 7A is modified from Linde (The Linde Group)[3].

---

[3] see http://muniche.linde.com/international/web/lg/spg/like35lgspg.nsf/docbyalias/anal_hplc.

5.2) Materials

1. micropipettes of 200–1000 microlitres (µl) and tips
2. 1.5 millilitre (ml) Eppendorf tubes
3. microcentrifuge (one per class or one per group)
4. sterile distilled water
5. vortex (one per class or one per group)
6. plastic pistil for grinding tissue
7. liquid nitrogen (−197 °C!!!)
8. 60% (v/v) acetone: 40% (v/v) ethyl acetate
9. HPLC vials and liners

5.3) Procedure

5.3.1) Total pigment extraction for HPLC analysis

Using the same plants as per Activity 3, use one set of plants (wild type and mutant) to extract plant pigments for HPLC analyses. This method is adapted from the one described in (Förster *et al.* 2009).

IMPORTANT

YOU WILL BE PROVIDED WITH CHEMICAL HAZARD INFORMATION. PLEASE READ THE RELEVANT SHEETS BEFORE STARTING THE EXPERIMENT. SEE APPENDIX G. ALWAYS WEAR SAFETY GOGGLES AND GLOVES WHEN WORKING WITH LIQUID NITROGEN AND ORGANIC SOLVENTS THROUGHOUT THE ACTIVITIES IN THIS MANUAL.

Materials	Method
Eppendorf tubes	1. Harvest three x 30 milligrams of leaf material from your wild type and mutant plant.
Liquid nitrogen	2. Freeze the material immediately by dipping the sample in liquid nitrogen. IMPORTANT: wear goggles and gloves to avoid burning your skin. Liquid nitrogen boils at −196 °C (very cold!) and causes rapid freezing when in contact with living tissue.
Pistil for grinding	3. Grind the frozen tissue into a paste.
Acetone/ethyl acetate, pipette, tips	4. Add 500 µl of filtered 60% (v/v) acetone: 40% (v/v) ethyl acetate to your ground tissue samples.
Vortex	5. Vortex for ten seconds.
H_2O, vortex, microcentrifuge	6. Add 400 µl of H_2O to each sample.
Vortex	7. Vortex for ten seconds.
Centrifuge	8. Centrifuge for five minutes at 16,000 g (depending on rotor, approx. 13,000 revolutions per minute (rpm))
HPLC vials	9. Label HPLC vials and place the bottom inserts in.

Materials	Method
Eppendorf tubes	10. Recover approximately 200 µl of the upper phase containing the pigments into another labelled tube. It is no problem if some of the lower phase is also carried over.
Microcentrifuge	11. Spin the pigment at 13,000 rpm for three minutes.
	12. Transfer 100 µl to the corresponding HPLC vial bottom insert.

5.3.2) HPLC run

This step will be performed by your demonstrator using an HPLC system. This method is adapted from the one described in (Förster *et al.* 2009).

You will be invited for a tour and explanation of the technique.

Materials	Method
	1. The demonstrator will load 10 µl of the extract onto the HPLC.
Waters Spherisorb 5 micrometre (µm) ODS2 column for reverse-phase HPLC (Agilent Technologies HP1100 series); acetonitrile: water: triethylamine, 90:10:0.1, v/v; ethyl acetate	2. Separate pigments using a linear gradient decreasing solvent A (acetonitrile: water: triethylamine, 90:10:0.1, v/v) from 100% to 33% (v/v) while increasing solvent B (ethyl acetate) from 0% to 67% (v/v) over 31 minutes, followed by a four-minute elution with 100% (v/v) solvent B at a flow rate of 1 mL min–1.
	3. Your instructors will give you the raw data and explain how to interpret it. See also Fig. 7 for an example.

5.4) Expected outcomes

1. Graph Abs vs retention time.
2. Identify peaks using the example provided in Fig. 7B. Specifically, look at potential differences in peak composition (retention time) between wild type and mutant samples.

Activity 6: Quantification of anthocyanins

6.1) Introduction and objectives

Anthocyanins are a subset of a large group of plant phenolics called flavonoids. These water-soluble pigments accumulate in the vacuole and play protective roles during acclimation to high light. Specifically, anthocyanins are glycosylated flavonoids and are responsible for the pink, purple and blue colours in plant organs (Fig. 5) (Lee *et al.* 2008). The colours are determined by the chemical groups attached to the ring B of the basic flavonol skeleton and the pH of the vacuole. Anthocyanins accumulate during stress conditions for example, such as a lack of nitrogen. Thus the presence or absence of anthocyanins can be used as an indicator for stress responses within the plant.

You will extract total pigment from frozen, ground leaf tissue from four-week-old wild type and mutant plants. You will then measure absorbance, quantify anthocyanins, and analyse the spectrum of the extract in the range of 400–800 nanometres (nm).

The main objectives of Activity 6 are to:

1. isolate total anthocyanins from both wild type and mutant plants grown in the light using a simple organic solvent extraction procedure
2. quantify anthocyanins using their spectrophotometric properties.

IMPLEMENTING GOOD DESIGN PRINCIPLES IN ACTIVITY 3-6:

PLANT DETECTIVE TIP

1. REPLICATION AND SAMPLING BIAS: CAREFULLY CONSIDER YOUR SELECTION OF LEAF SAMPLES. BULKING ACROSS LEAVES MAY BE NECESSARY, BUT ENSURE THAT YOU HAVE A RANDOM SELECTION OF LEAVES SO YOU ARE CERTAIN THAT EACH SAMPLE IS AN INDEPENDENT BIOLOGICAL REPLICATE. HOW CAN YOU ACHIEVE THAT?

2. REPLICATION: SOME OF THE EXPERIMENTS IN THIS ACTIVITY WILL HAVE A SAMPLE OF ONE FOR LOGISTICAL REASONS (N=1). REMEMBER THAT, WHILE NOT MUCH CAN BE EXTRAPOLATED FROM A SAMPLE OF ONE, THESE RESULTS CAN BE USED TO SUPPORT OTHER DATA.

3. DATA TRAIL: RECORD FROM WHICH PLANTS YOU TOOK EACH OF YOUR SAMPLES.

4. BLINDING: BE AWARE OF EXPERIMENTER BIAS WHEN TAKING MEASUREMENTS.

6.2) Materials

1. 1% (v/v) HCl
2. 1.5 millilitre (ml) Eppendorf tubes
3. 70% (v/v) methanol
4. centrifuge (tabletop or microcentrifuge)
5. chemical hood
6. chloroform
7. gloves
8. goggles, one per person
9. liquid nitrogen (−197 °C!!!)
10. MilliQ water
11. P1000 and P200 micropipettes, and tips (one each)
12. pestle for grinding; alternatively, a mortar and pestle could be used
13. plate reader BIO-TEK uQuant and 96 well plate or Spectrophotometer and cuvette
14. vortex (one per class or one per group)

6.3) Procedure

Using the protocol below (based on published methods by (Lee *et al*. 2005) and (Neff and Chory 1998)) you will extract anthocyanins and quantify their content in your leaf tissues. Note that teaching staff will perform the harvesting. Depending on the type of grinding method, you will be asked to continue from Step 3 or Step 4.

6.3.1) Anthocyanin extraction

Materials	Method
Eppendorf tubes	1. Harvest approximately 30 milligrams of leaf tissue from FOUR wild type and mutant plants and place each into a 1.5 ml Eppendorf tube. Make sure tubes are properly labelled beforehand. TIP: label tubes with your group number, genotype, and replicate. For example, 01–WT–02 is the replicate number two of the wild type leaf tissue of Group 1.
Goggles, gloves, liquid nitrogen	2. Freeze the material immediately by dipping the sample in liquid nitrogen. IMPORTANT: wear goggles and gloves to avoid burning your skin. Liquid nitrogen boils at −196 °C (very cold!) and causes rapid freezing when in contact with living tissue. Your eyes are very vulnerable to liquid nitrogen. EXERCISE CARE and consider your surroundings and anyone nearby.
Pestle for grinding	3. Grind the frozen tissue.
Pipettes and tips, methanol, vortex	4. Add 400 microlitres (µl) of acidified methanol (70% methanol; 1% (v/v) HCl) to your ground samples.
Vortex	5. Vortex the samples *vigorously* for 30 seconds.
Fume hood, chloroform, water, vortex	6. Add 700 µl of chloroform and 300 µl of H_2O to the sample.
Vortex	7. Vortex the samples for ten seconds.

Materials	Method
Centrifuge	8. Centrifuge at full speed (13,000 revolutions per minute) for five minutes.[4]

6.3.2) Anthocyanin quantification

Materials	Method
Eppendorf tubes	1. Transfer as much as possible (~500 µl) of the upper (aqueous) phase to a new tube. If you are not sure about the recovered volume, you can estimate it using a pipette. Ask your demonstrator.
Plate reader or spectrophotometer, plate or cuvette.	2. Use 200 µl of the 'blank' solution[5] supplied by the plate reader operator along with your samples to read absorbance using the plate reader.
	3. Measure sample absorbance at 530 and 657 nm. The operator will enter the individual absorbance readings into an Excel spreadsheet.
	4. Calculate the anthocyanin concentration on a tissue gram basis as: $$[Anthocyanin](mg\,/\,gFW) = \frac{(Abs_{530} - Abs_{657}) \times MW \times DF \times V}{\varepsilon \times l \times FW}$$ Where: ε = the molar absorptivity constant of cyanidine–3–glucoside (26900 L cm^{-1} mol^{-1}); monomeric anthocyanins can be expressed as cyanidine–3–glucoside. l = path length: one centimetre (cm); however this may need to be recalculated **when using the plate reader**. MW: molecular weight of the cyanidine–3–glucoside is 449.2 g mol^{-1} DF: Assuming that 500 µl were extracted in Step 9, and that *most* of the anthocyanin fractionated in the methanol–water fraction (400 µl + 300 µl), *DF* equals 700/500=1.4. *Volume* = volume of the aqueous phase extracted on Step 4 in ml (0.4 ml) *FW* = tissue mass in grams (g)
	5. If the colour of your extract is reddish, then take 200 µl to a new tube and add 10 µl of 1N NaOH. Observe what happens with the colour. If nothing happens add more NaOH. You can do this for one or both samples.

6.4) Expected outcomes

1. Calculate the amount of anthocyanins in both genotypes in micrograms (µg) of pigment per gram (g) of tissue (µg g^{-1}).

2. Also, note whether there are any obvious differences in 'reddish' colour in the rosette leaves, or under the rosette, between different genotypes?

3. Is there a noticeable stress trait that can be correlated with the anthocyanin content (i.e., a leaf that is more wilted has a different anthocyanin content to one that is not)?

---

[4] By this stage any samples with anthocyanin levels significantly higher than control tissue should be at least faintly pink

[5] The blank solution was prepared exactly as per instruction, but without ground tissue.

Activity 7: Gas exchange measurements on intact leaves — photosynthetic responses to light and CO_2

7.1) Introduction and objectives

It is possible to determine rates of CO_2 assimilation and water loss (transpiration) by measuring the flux of CO_2 and water vapour from a leaf in a sealed chamber. This process, termed gas exchange (because CO_2 is going in and water vapour is coming out) is more complicated than might be initially imagined. During photosynthesis, plants take up CO_2 (which is converted to sugar) and produce oxygen. All the while they are respiring and releasing CO_2 back into the cells. To make matters more complex, the enzyme that fixes CO_2 (Rubisco) also 'fixes' oxygen, a reaction called photorespiration that releases CO_2, but does not produce energy. Gas exchange is also influenced by light levels, because when more light is available, generally, more CO_2 can be fixed.

In this activity you will measure gas exchange on your wild type and mutant plants and determine whether the plants differ in rates of carbon fixation or water loss at a set of standardised conditions. Your photosynthesis measurements will be done using an infrared gas analyser (IRGA) (Fig. 8). The IRGA uses infrared radiation to detect the concentration of H_2O and CO_2 in the air being pumped over a leaf in a chamber. In effect, the gas concentrations are compared before and after being passed by the leaf. Using information about the change in H_2O and CO_2, the IRGA calculates photosynthesis and stomatal conductance of water. Consider for a moment a well-lit leaf in the chamber — how would you expect the gas concentrations to change before and after exposure to the leaf? What if the chamber was darkened?

Examining the shape of the photosynthetic response to light reveals several important things about the biology of a leaf (Fig. 8C). In the dark, no photosynthesis takes place in C_3 leaves and respiration (production of energy using O_2 and producing CO_2, just like animals do) is greater than photosynthesis. Therefore, plants show a net production of CO_2. When light levels rise, photosynthesis starts and CO_2 is taken up and fixed. The higher the light, the more CO_2 can be fixed, up to a certain point when CO_2 uptake is saturated. By determining the response of photosynthetic rate to light, we can identify the following parameters:

1. R_d: respiration rate
2. LCP: the light compensation point, or the light level at which respiration and photosynthesis balance each other
3. LSP: the light saturation point, or the light level at which photosynthetic rate ceases to increase
4. Φ: quantum yield
5. photosynthetic capacity, or maximum rate of assimilation
6. transpiration rate (T): although not a photosynthetic parameter, T is an estimation of the amount of water leaving the leaf per unit area per time
7. stomatal conductance (g_s): an estimation of how open stomatal pores are to allow flux of both water vapour and CO_2 out and in the leaf

If time permits you may also be able to measure a CO_2 response, or 'A vs. C_i' (AC_i), where A is the photosynthetic rate and C_i the CO_2 concentration inside the leaf. An AC_i curve measures the rate of CO_2 assimilation, generally at high light levels, and across varying internal CO_2 concentrations. We use these curves to indicate the maximal photosynthetic capacity of the leaf in the absence of limitation by CO_2 concentration. These curves can also be used to indicate electron transport capacity and characteristics of Rubisco kinetics. Your demonstrator will help you to measure and interpret an AC_i curve.

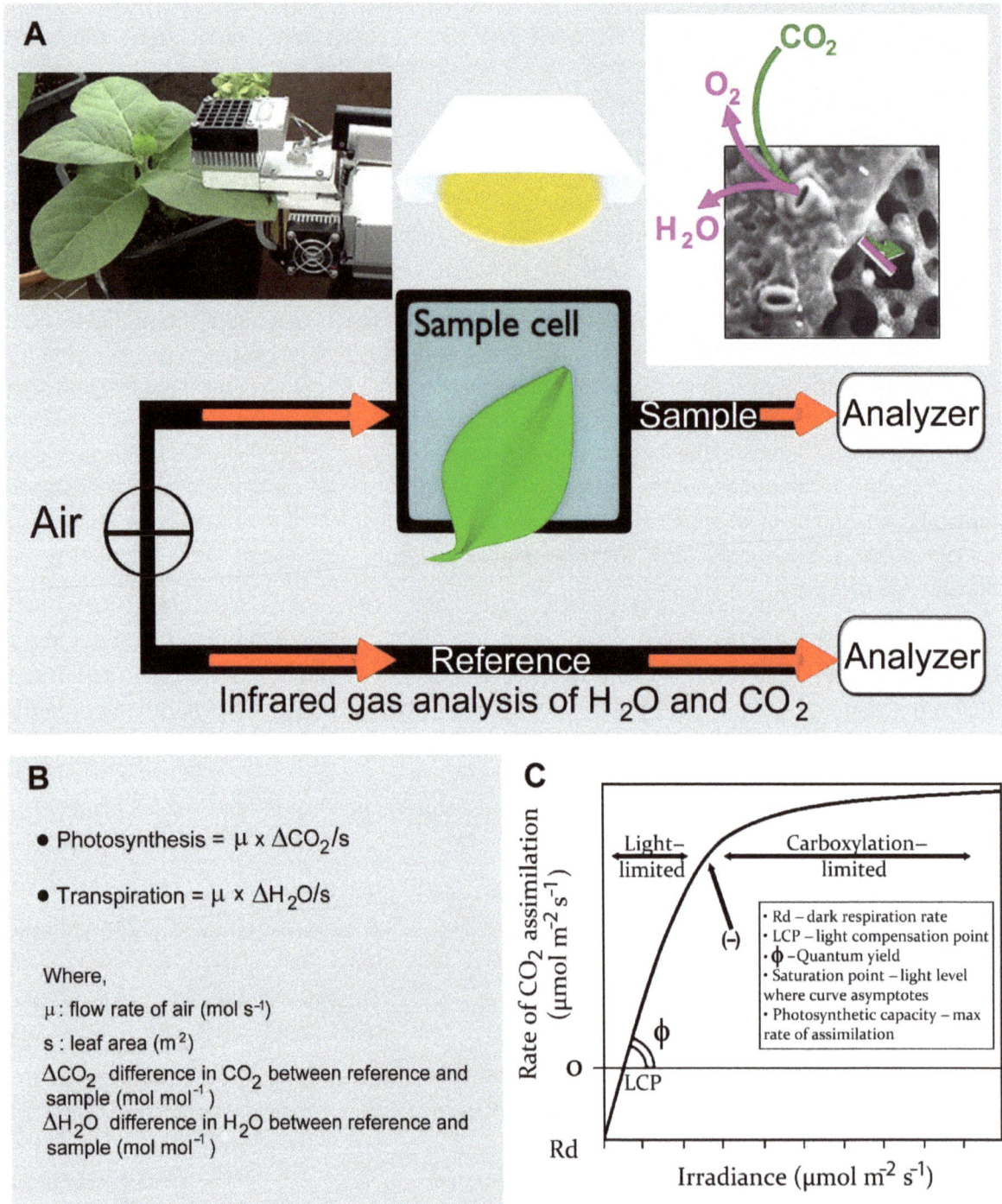

Figure 8. Basics of gas exchange measurements and the light curve
A) Schematic representation of the gas exchange system LiCor. B) Basic equation for photosynthesis and transpiration. C) Light curve resulting from plotting rate of CO_2 assimilation versus increasing light intensity (irradiance). Note that assimilation is first limited by the amount of light and then by the rate of carboxylation and recycling of the required precursors. Schemes courtesy of Susanne von Caemmerer, The Australian National University.

If you are running out of time, you may want to measure assimilation only at one standard condition of ambient CO_2 and high light and replicate these measurements to compare your mutant and wild type.

A porometer will be used to measure stomatal conductance (g_s). The porometer has a time advantage over the IRGA, though both can tell you about plant water use. The porometer is a more simple piece of equipment that measures temperature and water content in the air around the leaf using a small, lightweight chamber. Because the porometer is quicker to use, more measurements of g_s in the abaxial side of the leaf (bottom side) can be made resulting in a large sample of data being available for statistical analyses.

Gas exchange is very sensitive to measurement conditions, including light level, temperature, CO_2 concentration, water vapour/relative humidity in the chamber and, perhaps most surprisingly, time. When taking your measurements, record information about the measurement conditions. Also, remember that the porometer, unlike the IRGA does not have a light source. Consider the effect of light on stomata and make sure that your plants are in an appropriate light environment before performing the porometer measurements. Finally, be aware that plants often show a reduction in photosynthetic rates as the day progresses, sometimes 'shutting-down' entirely by mid- to late afternoon. We recommend that all gas exchange measurements are done early in the day. Alternatively, your instructor may grow your plants in a growth chamber with the timing of daybreak shifted relative to real time, so that your plants behave as if it is morning even when assayed in the afternoon.

The objectives of Activity 7 are to:

1. become familiar with a portable gas exchange system to determine the photosynthetic rates (A). Our protocol is based on a Li-Cor 6400 model IRGA, but can be adapted to any other instrument

2. compare total photosynthesis and stomatal conductance between wild type and mutant plants

3. run a 'light curve' for the wild type plant and, if time permits, for the mutant plant.

IMPLEMENTING GOOD DESIGN PRINCIPLES IN ACTIVITY 7:

PLANT DETECTIVE TIP

1. **REPLICATION:** SOME OF THE EXPERIMENTS IN THIS ACTIVITY WILL HAVE A SAMPLE OF ONE (N=1) FROM WHICH NOT MUCH CAN BE EXTRAPOLATED. IN OTHER EXPERIMENTS YOU CAN HAVE REAL BIOLOGICAL REPLICATES. MAKE THE MOST OF THOSE OPPORTUNITIES.

2. **DATA TRAIL:** FAMILIARISE YOURSELF WITH THE VARIABLES THAT YOU WILL RECORD IN YOUR SPREADSHEETS. MAKE SURE YOU KNOW WHAT EACH MEANS AND THAT YOU HAVE RECORDS OF ALL THE MEASUREMENT CONDITIONS, PLANT ID ETC.

3. **BLINDING:** CONTROL FOR EXPERIMENTER BIAS.

7.2) Materials

1. Li-Cor gas exchange system
2. CO_2 cartridges
3. desiccant
4. charged batteries for Li-Cor or adaptor for main power

7.3) Procedure

7.3.1) Measuring gas exchange parameters using an IRGA

It is likely that your instructor will operate the IRGA equipment for you. This is fine because it is a complicated piece of equipment that takes practice and experience to use. In the Plant Detectives Project we are more interested in you learning the principles than how to run the machine[6].

The measurements will demonstrate the dependence of photosynthetic processes on measurement conditions and growth conditions, as well as allowing you to assess any differences due to the genetic make-up of the study plant.

Your instructor will aim to complete the following with you over the course of the exercise:

1. a light response curve for a single wild type and mutant plant

---

[6] To learn more about the specifics of the machinery and the measurements, visit http://prometheuswiki.publish.csiro.au/tiki-index.php?page=Gas+exchange+measurements+of+photosynthetic+response+curves.

2. if time permits, measure photosynthesis at saturating light for two more replicates of the wild type and mutant. By combining your data with that from the light response curve from the same light level you will have three replicates to use in statistical analysis. If time permits you can measure an AC_i (photosynthesis vs. CO_2 concentration) curve on a single wild type and mutant plant to get an idea of how CO_2 concentration affects photosynthetic rate and water loss.

Record the settings used for the measurement conditions in your laboratory notebook. We have previously found that the conditions in parentheses below work well for measurements of Arabidopsis plants grown in growth chambers or controlled conditions in a glasshouse:

1. [CO_2] reference: the CO_2 level at which measurements are being made (400 parts per million (ppm))
2. flow: how much the air is circulating in the system (300 ml min^{-1})
3. Temperature: the temperature inside the measuring chamber (22 °C)
4. PAR (µmol m^{-2} s^{-1}): photosynthetically active radiation. For the light curve use in this order: 800, 500, 350, 100, 50 and 0.
5. RH: relative humidity (or VPDI or [H_2O]) — how much water is in the air. Aim to maintain between 50–60%

Record the following information into a table in your notebook or an Excel spreadsheet for each measurement you take. Your instructor may also provide you with a spreadsheet of all the data produced by the IRGA, but you'll find the key measures are:

1. plant number and whether wild type or mutant
2. CO_2 R: CO_2 concentration in the reference cell (i.e., before passing over the leaf) — this is set by the operator
3. PAR: this is set by operator
4. temperature of the block: this is set by operator
5. RH: this varies, but the operator aims to keep it between 50 and 60%
6. photo: photosynthetic rate in terms of uptake of CO_2, µmol m^{-2} s^{-1} (note, in the dark this will be negative, respiration releases CO_2)
7. cond (g_s): rate of stomatal conductance to water
8. C_i: internal concentration of CO_2, ppm

Sample data sheet — fill all cells in the spreadsheet, even if it means repeating some information on subsequent lines (e.g., genotype and plant number will repeat over every line of a given set of light response curve measurements)

Genotype	Plant Nº	[CO_2]Ref	PAR	Photo	Conductance (g_s)	C_i	Comments
		(ppm)	(µmol m^{-2} s^{-1})	(µmol m^{-2} s^{-1})	(mol m^{-2} s^{-1})	(ppm)	

Genotype	Plant Nº	[CO$_2$]Ref	PAR	Photo	Conductance (g$_s$)	C$_i$	Comments

7.4) Expected outcomes

1. You will be able to plot A (photosynthetic rate) vs light intensity and g_s (stomatal conductance) vs light intensity for the individual plants measured. Make these plots and draw a curve by hand or using software if you prefer. Based on your light-response curve, determine what light level is needed to achieve ~90% of the maximum photosynthetic rate (e.g., where does the rate start levelling off?).

2. Using the curve, approximate the compensation point, saturation point, quantum yield, and light-saturated photosynthetic rate for your wild type and mutant plants. Do they look very different?

3. At saturating light, is there any difference in A or g_s between the wild type and mutant plants? (If you managed to take some extra measurements you will be able to assess this statistically.)

4. If you were able to measure it, what is the maximum photosynthetic capacity for your wild type and mutant plants in the absence of stomatal (CO$_2$) limitation? How does this compare to the measurements at ambient CO$_2$ levels?

Activity 8: Measuring stomatal conductance to water using a porometer

8.1) Introduction and objectives

Stomata confer a physical barrier to water vapour diffusion. The conductance to water vapour is a major determinant to water loss and availability of CO_2 for photosynthesis. Although both CO_2 and water vapour can be quantified by using an infrared gas analyser (like the LiCor, Activity 7), another instrument, the porometer (Fig. 9), can provide reliable estimates of stomatal conductance. The cycling porometer works by measuring the time it takes for a leaf to release enough water vapour to change the relative humidity in a small chamber by a fixed amount. By comparing this amount of time with a calibration plate, the stomatal conductance can be calculated (Monteith *et al.* 1988).

Unlike with the LiCor, environmental conditions, such as light, temperature and relative humidity, cannot be controlled in the porometer chamber. The porometer, however, is easy to calibrate and use, and it provides readings much faster than the LiCor, making it the tool of choice when sampling many plants.

The objectives of Activity 8 are to:

1. become familiar with the use of the porometer
2. quantify and compare stomatal conductance between wild type and mutant plants.

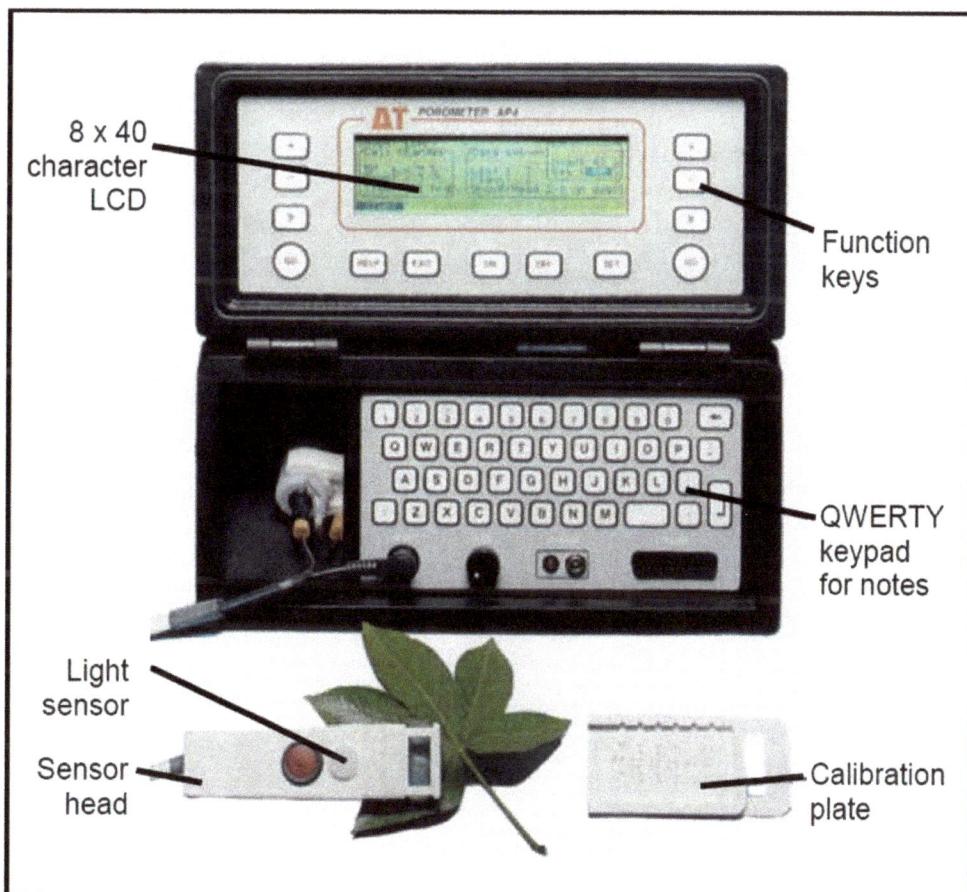

Figure 9. The porometer
Figure extracted from brochure 'AP4 Porometer Data Sheet' (http://www.delta-t.co.uk/product-support-material.asp).

IMPLEMENTING GOOD DESIGN PRINCIPLES IN ACTIVITY 8:

PLANT DETECTIVE TIP

1. **REPLICATION** AND SAMPLING BIAS: CONSIDER CAREFULLY HOW YOU SELECT THE PLANTS ON WHICH YOU MEASURE CONDUCTANCE.

2. **DATA TRAIL:** RECORD FROM WHICH PLANTS YOU TOOK EACH OF YOUR SAMPLES.

3. **BLINDING:** REMEMBER EXPERIMENTER BIAS WHEN TAKING YOUR MEASUREMENTS.

8.2) Materials

1. porometer type AP4, Delta-T
2. porometer calibration plate and plastic ziplock bag
3. Whatman paper, pre-cut to fit into calibration plate
4. plastic bag
5. water
6. dry desiccant

8.3) Procedure

Porometer datasheets and manual: Delta-T Devices.

8.3.1) Porometer calibration

The AP4 is supplied with a moulded polypropylene calibration plate with six groups of holes; the rate of diffusion of water vapour through these holes has been carefully verified. Water vapour is provided by backing the plate with dampened paper. The sensor head is clipped onto the calibration plate, and readings are stored from each of the six standard calibration positions.

Materials	Method
Calibration plate, damp filter paper	1. Place damp filter on the back of the calibration plate to cover all holes. Seal with sticky tape. Make sure to remove air pockets before sealing
Plastic bag	2. Place the calibration plate inside the ziplock bag and keep it at the same temperature as the location at which the measurements will be run. TIP: it is best to prepare the calibration plate on the day before running the experiment.

Materials	Method
Porometer	3. Select Calibration from main menu.
	4. Select RH and press + or – to disable the pump; open the head and wave it about to determine the ambient RH. Adjust the Set RH value to the closest to that of ambient RH. Press Set to change other values (like the cup type, or units) or Go or Exit to return to the Insert Plate screen.
	5. Place the head in the corresponding position of the calibration plate that is indicated in the Position column and press Go to start. Repeat this step at the corresponding position when prompted by the software.
	6. After the six positions are completed, press Fit Curve. Accept if error is <5%. NOTE: you can come back to individual positions if there is a major discrepancy between the theoretical and estimated values.

8.3.2) Porometer measurement

Clamp a leaf with the measuring head and press Go (NOTE: the red button in the measuring head). Attempt to measure three leaves per plant, five plants per genotype.

8.4) Expected outcomes

1. Using the data from the porometer compare conductance measurements for your wild type and mutant plants. See Appendix B for information on how to assess whether there are statistically significant differences in conductance between your wild type and mutant plants.

Activity 9: Drought response

9.1) Introduction and objectives

The previous activities have compared the structure and function of plants under normal growth conditions. Sometimes the effects of mutations are not apparent unless a plant is challenged in some way. By following the protocol below, you will analyse the response of both wild type and mutant plants to drought stress. In addition to following the growth and development of you plants during and after the drought, you will determine the relative water content of your plants.

The major goals of Activity 9 are to:

1. investigate the effect of water deficit on wild type plants
2. to compare these to the effects on the mutant.

You could also perform some of the assays from the previous activities if you think they can provide useful information. Discuss your interests with your peer mentor and instructor.

FACTORIAL DESIGNS:

WHEN YOU INTRODUCE A DROUGHT (OR SUBSEQUENTLY ABA) TREATMENT ON YOUR PLANTS YOU WILL EXPLORE NEW DESIGN TERRITORY. YOU HAVE JUST ENTERED THE WORLD OF FACTORIAL DESIGNS. WELCOME. A FACTORIAL DESIGN IS ONE WHERE YOU HAVE MORE THAN ONE FACTOR. FACTORS CAN BE MULTIPLE TREATMENTS, OR IN OUR CASE THEY ARE GENOTYPES AND TREATMENTS. WE WANT TO COMPARE THE EFFECT OF THE TREATMENT ON THE TWO GENOTYPES AND ASSESS WHETHER THE GENOTYPES RESPOND IN SIMILAR OR DIFFERENT WAYS. TO DO THIS WE MUST RANDOMLY ASSIGN OUR GROUPS OF PLANTS WITHIN GENOTYPE TO THE TWO DIFFERENT TREATMENTS. IDEALLY WE WOULD ARRANGE OUR TRAYS SO THAT WE HAD MULTIPLE BLOCKS WITH REPRESENTATIVES OF EACH FACTORIAL COMBINATION IN EACH. FOR LOGISTICAL REASONS THIS MAY BE DIFFICULT IN YOUR EXPERIMENT. BUT, YOU SHOULD CONSIDER THE IMPLICATIONS OF THESE DESIGN LIMITATIONS. DISCUSS AND BE ABLE TO EXPLAIN.

9.2) Materials

1. plants in pots: in general, plants between four to five weeks old are optimal for the drought experiment. The growth rate may vary, however, among batches of plants from different years and in different growing conditions. Teaching staff will advise you as to which of your plants (first or second round of sowing) are best for this assay
2. balance (one–two per class)
3. aluminium foil

4. plastic Petri dishes (20)
5. permanent marker (one)
6. oven 60 ºC to 80 ºC
7. Whatman filter paper (few pieces)
8. scissors

IMPLEMENTING GOOD DESIGN PRINCIPLES IN ACTIVITY 9:

PLANT DETECTIVE TIP

1. **DESIGN:** WHEN YOU INTRODUCE YOUR DROUGHT TREATMENT YOU ARE USING WHAT IS CALLED A FACTORIAL DESIGN (SEE BOX ABOVE), KEEP THIS IN MIND AS YOU ARRANGE YOUR PLANTS AND ALLOCATE YOUR TREATMENTS.

2. **REPLICATION:** MAKE SURE YOU HAVE ENOUGH BIOLOGICAL REPLICATES TO GET A MEAN AND MEASURE OF VARIABILITY.

3. **BLOCKING AND BLINDING:** CONSIDER THE ORDER IN WHICH YOU TAKE THE MEASUREMENTS AND EXPERIMENTER BIAS.

DATA TRAIL: MAKE SURE TO RECORD YOUR DATA.

9.3) Procedure

9.3.1) Experiment set up

Materials	Method
Trays, plants	1. Move at least six plants of each genotype into two separate trays. Randomly place a combination of the wild type and mutant plants in each tray (i.e., don't have half on one side of the tray and half on the other). Remaining plants can be used for other assays.
Labels, marker	2. Label one tray as 'WS–Drought; do not water'. Label the other tray as well-watered 'WW–Control'.
	3. The drought treatment will be initiated for you seven days before the practical, when teaching staff will water the trays with 200 millilitres (ml) of water instead of 800 ml. You will start your observations on day seven of the drought, but you will be able to check your plants during lunch on each day in the following week. Depending on the performance of the plants, you may need to continue observations and measurements until the end of the course.

Materials	Method
	4. Observe the growth and development of wild type and mutant plants after the drought treatment is imposed. Note any differences in phenotype between these two genotypes during water-deficit conditions.

9.3.2) Relative water content determination

When measuring plant water content, raw water content is not useful for comparison between plants with different morphologies. Instead, this information can be normalised by determining the relative water content (RWC) as:

$$RWC = \frac{Fw - Dw}{Tw - Dw}$$

Where Fw is the fresh weight at harvest, Dw is the dry weight of the sample, and Tw is the turgid weight, or the maximum weight at the highest level of water holding capacity. On one day of the drought treatment (to be decided along with your supervisor), you will perform the determination of RWC as follows to check for differences in:

1. the effect of the growth condition on the water content of the wild type plant
2. the effect of the genotypes on the water content at both conditions.

Materials	Method
Scissors, plants	1. Detach the rosette from the roots, carefully removing traces of soil. Aim to use at least five individuals of each genotype under both control and drought conditions (total 20 plants). Make sure plants are labelled clearly to help organise the data organised. Consider design principles when selecting plants.
Scale	2. Record the Fw at time 0.
Plastic Petri dishes, H_2O	3. Place the rosettes in a Petri dish half-filled with water. Leave the rosette in the dish (incubate) for at least four hours at 4 ºC in the dark. You can also perform this incubation overnight and weigh the samples the next day.
Blot paper	4. Blot the leaves and record the Tw.
80 ºC oven	5. Place the rosettes in paper envelopes and dry at 80 ºC overnight (24–48 hours minimum). Make sure the envelopes are properly labelled with the genotype, plant number, and group.
	6. Weigh to determine the Dw at least 24 hours later.
	7. Calculate the RWC as (Jones 2007): $RWC = \frac{Fw - Dw}{Tw - Dw}$

9.3.3) Rosette dehydration experiment (optional)

The monitoring of water loss of detached rosettes over a short period of time provides information about dehydration mechanisms. This is easily done by detaching rosettes and weighing them regularly over time to determine the dehydration rate. By plotting percentage of original weight vs time it is possible to determine whether a leaf regulates water flux through stomatal control (the initial phase of water loss) or cuticular evapo-transpiration (the second phase of water loss). Both processes may differ between the wild type and mutant plants.

Materials	Method
Scissors, plants, scale	1. Cut a rosette from a well-watered (WW, control) plant and weigh it at regular intervals for two hours. The preferred intervals are: every 10 minutes for the first hour, and every 20 minutes for the second hour.
	2. Plot percentage of original weight vs time.

9.3.4) Gas exchange (optional)

If your plants are big enough (or still sufficiently alive) you may want to measure stomatal conductance during your drought experiment using the porometer, or LiCor if leaves are big enough and turgid. Consult with your supervisor. If possible, measure gas exchange on at least three plants per genotype per water treatment so that you can assess the effects of drought stress statistically. Refer to activities 7 and 8.

9.3.5) Harvest (optional)

Harvesting the above- and below-ground parts of your plants will enable you to do final growth measurements and determine leaf mass to area ratio (LMA) and root to shoot ratio. This optional activity can be performed on the well-watered plants or both.

Materials	Method
Plants, paper bags, labelling pen	1. Remove above-ground parts at soil surface for three–five plants each of wild type and mutant.
	2. You may like to count leaves at this stage or measure the area of some or all leaves (see below).
	3. Separate any reproductive structures (bolt, flowers and siliques) and place in a labelled paper bag.
	4. Place rosette in a labelled paper bag.
Rubbish bin, trays or buckets of water, sieves	5. Gently tip plant from pot and remove as much soil from roots as possible with your hands.
	6. Place roots in a tray of water and tease out all remaining soil. You may need to rinse the roots, or to change the water periodically. Make sure that soil does not go down the sink.
	7. Dispose of pots and soil as directed.
Drying oven, scales	8. Place bags in drying oven (24–48 hours minimum) and return to weigh all parts.

9.3.6) Measuring leaf areas and LMA (optional)

The area and shape of individual leaves can be measured using photographs or scans of leaves.

Materials	Method
Scanner or digital camera, leaves	1. Scan or photograph your leaves. Make sure leaves are labelled and each image has a scale marker on it.

Materials	Method
	2. Determine leaf area using ImageJ software[7].
Paper bag, labelling pen, scales	3. Place leaves in labelled paper bag, dry as for 9.3.2, weigh to calculate LMA.

9.4) Expected outcome

For all plants grown at both conditions:

1. Describe the phenotype of wild type and mutant plants grown under both conditions. When do plants start wilting? Are there pigmentation differences? Do the observations correlate with the findings in activities 4 and 5?

2. Calculate and plot the RWC. Are there statistically significant differences? Remember, this is a factorial design so statistical tests will need to be two-way tests. See Appendix B.

3. Are there any differences in final biomass (D_w)?

4. How is the g_s affected by drought? Are there any statistically significant differences?

5. From harvest data, compare the growth of your plants with and without drought. Do the plants differ in total mass? Does relative allocation above and below ground differ between genotypes? Between drought treatments?

6. Do leaf area and LMA differ?

---

[7] See http://prometheuswiki.publish.csiro.au/tiki-custom_home.php.

Activity 10: Microscopy analyses

10.1) Introduction and objectives

Thus far you have phenotyped your plants by comparing external morphology, architecture and growth. By following the protocol below, you will use basic microscopic techniques to look at plant anatomy. The activity focuses on stem anatomy, but you may also choose to look at leaf and root anatomy. Allow your plants to be your guide in this exercise: if you can see a difference in a structure between your wild type and mutant plants, dissect that structure and endeavour to determine what has changed.

The objectives of Activity 10 are to:

1. examine the epidermal morphology and vascular tissue
2. compare any changes in wild type and mutant plant stomata number and aperture (i.e., how wide the stomata can open).

IMPLEMENTING GOOD DESIGN PRINCIPLES
IN ACTIVITY 10:

PLANT DETECTIVE TIP

1. QUALITATIVE VS QUANTITATIVE DATA: (SEE BOX IN ACTIVITY 1) IN SEVERAL OF THESE PROCEDURES YOU WILL QUALITATIVELY DESCRIBE YOUR RESULTS AND SHOW PICTURES TO ILLUSTRATE THEM. CONSIDER HOW YOU CAN QUANTIFY YOUR RESULTS AS WELL. FOR EXAMPLE, CAN YOU COUNT THE NUMBER OF OPEN VS CLOSED STOMATA IN A SET AREA? HOW WILL YOU GET TECHNICAL AND BIOLOGICAL REPLICATES OF THOSE COUNTS? CAN YOU MEASURE DIFFERENCES IN CELL SIZES OR ARRANGEMENT IN YOUR IMAGES?

2. REPLICATION: IF YOU GET TECHNICAL REPLICATION OF STRUCTURE WITHIN A LEAF OR STEM, REMEMBER THAT YOU NEED REPLICATION BETWEEN LEAVES OR STEMS AS WELL.

10.2) Materials

1. microscope slides
2. microscope (one)
3. sticky tape (Sellotape)
4. Petri dishes
5. vibratome
6. thin forceps (one–two)
7. razor blades (one–two)

10.3) Procedure

10.3.1) Examination of stem cross-sections

We will section stem pieces of Arabidopsis (of wild type and your mutant) using a vibratome. You will embed five-millimetre-long pieces of stems in 3% (w/v) agarose (provided) and make cross-sections of the stems. The use of the vibratome will be demonstrated in the class. The vibratome sections will be approximately 0.1 mm in thickness.

Transfer the sections to a glass slide and observe them under a light microscope. Compare the anatomy of the cross-sections of wild type and mutant; e.g., for total stem diameter, cell types and cell sizes, number of cell layers (Fig. 10).

10.3.2) Examination of root cross-sections (optional)

If your group has time and interest, you may also want to section root tissue. Ask your demonstrator for advice and then follow protocols described for stem tissue above. Note this may be tricky with thin Arabidopsis roots unless it has undergone secondary thickening and has a larger diameter than it did when it was a germinated seedling. Try this if you expect a root phenotype in your mutant.

10.4) Expected outcome

1. Describe any differences of the stem cross-sections between phenotypes. Do stems of the wild type and mutant differ in size or structure? A drawing is a good way to illustrate the morphology of stem cross-sections. Label all structures that you can identify. Record the size of anatomical elements and describe other differences, if any (record the scale and show it on your drawing). Alternatively the stem cross sections can be photographed and put in your laboratory notebook.

2. If it was observed, describe root cross-sectional anatomy and compare the two phenotypes. Do roots differ in thickness? Cell sizes? Arrangement of vasculature? Again, a drawing in your laboratory notebook is a good way to illustrate the morphology of cross-sections. Label all structures you can. Record size of anatomical elements and describe other differences, if any (record the scale and show on your drawing).

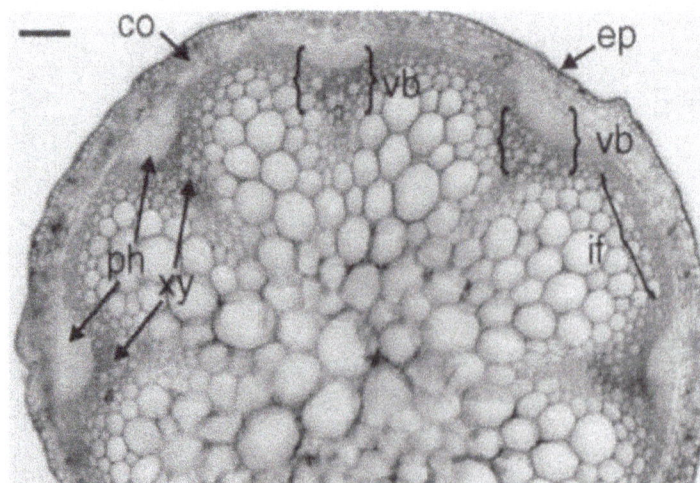

Figure 10: Cross-section of the stem from Arabidopsis
Micrograph showing a cross-section of Arabidopsis stem. vb: vascular bundles, xy: xylem, ph: phloem, co: cortex, ep: epidermis, if: intervascular region. Image from Turner and Somerville 1997.

Activity 11: Stomata and the effect of the hormone ABA

11.1) Introduction and objectives

In this activity we will look at the effect of the hormone abscisic acid (ABA) in promoting stomata closure. Stomatal response to the environment allows plants to control the loss of water (by closing stomata down) or intake of CO_2 (by opening stomata). ABA is produced under abiotic stress conditions, such as high light, water deficit or osmotic shock. One of the many effects of ABA is the promotion of the reduction of guard cell volume, thus inducing the closure of the stomatal pore. It is possible to artificially treat leaf peels containing both epidermal and guard cells with ABA and study the hormone's effect on stomata closure (Fig. 11). The protocol described below may be conducted in conjunction with the microscopy exercises described in Activity 10, or on its own.

The objectives of Activity 11 are to:

1. visualise and identify epidermal and guard cells
2. compare the response of the wild type and mutant stomata to ABA.

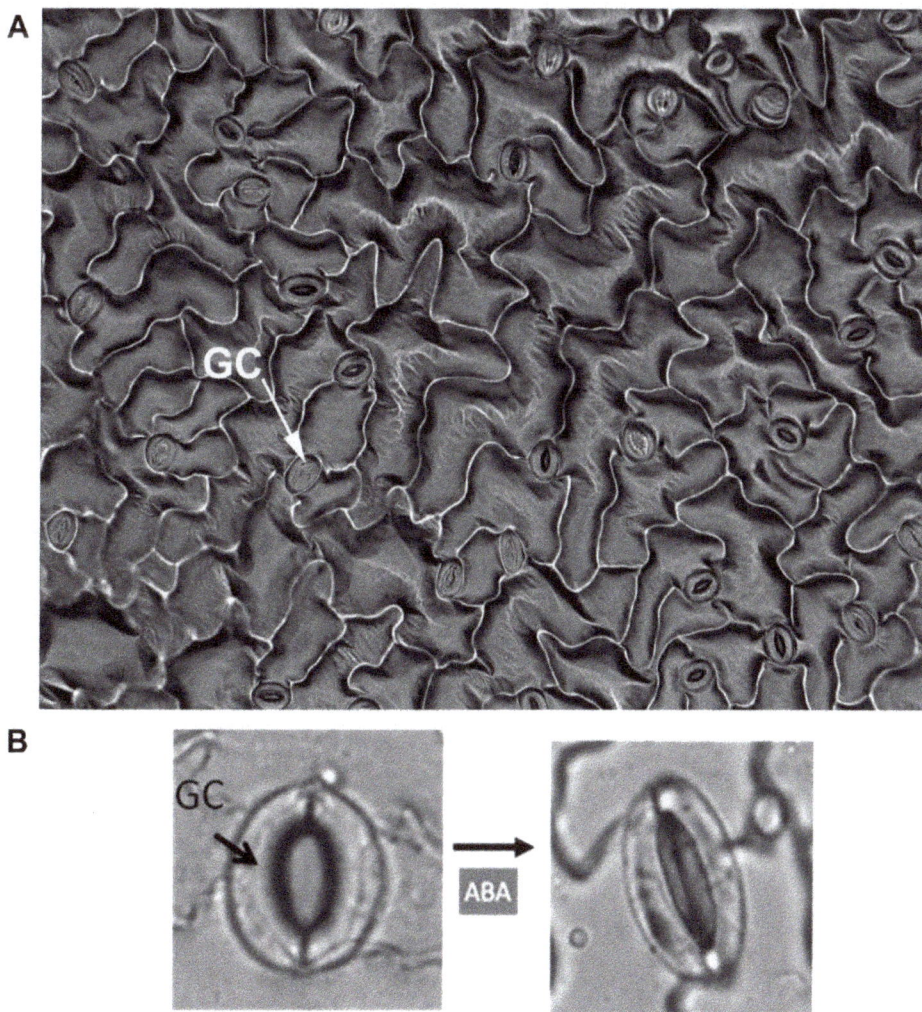

Figure 11. Micrograph of epidermis and stomata in Arabidopsis
A) Surface view of the Arabidopsis epidermis showing irregularly shaped epidermal cells and guard cells and the guard cells (GC) outlining the stomatal pore. B) Close up of the guard cells (GC) surrounding the pore. Incubation with ABA promotes stomata closure. Image courtesy of Nok Pornsiriwong (The Australian National University).

IMPLEMENTING GOOD DESIGN PRINCIPLES IN ACTIVITY 11:

PLANT DETECTIVE TIP

1. **QUALITATIVE VS QUANTITATIVE DATA:** (SEE BOX IN ACTIVITY 1) IN SEVERAL OF THESE PROCEDURES YOU WILL QUALITATIVELY DESCRIBE YOUR RESULTS AND SHOW PICTURES TO ILLUSTRATE THEM. CONSIDER HOW YOU CAN QUANTIFY YOUR RESULTS AS WELL. FOR EXAMPLE, CAN YOU COUNT THE NUMBER OF OPEN VS CLOSED STOMATA IN A SET AREA? HOW WILL YOU GET TECHNICAL AND BIOLOGICAL REPLICATES OF THOSE COUNTS? CAN YOU MEASURE DIFFERENCES IN CELL SIZES OR ARRANGEMENT IN YOUR IMAGES?

2. **BLOCKING:** WHEN YOU SET UP YOUR DISHES FOR THE ABA PROCEDURE AND ALLOCATE LEAVES TO DISHES CONSIDER YOUR DESIGN PRINCIPLES. CAN YOU CREATE BLOCKS OR ALLOCATE LEAF TISSUE FROM A GIVEN PLANT ACROSS THE TWO TREATMENTS? NOTICE, YOU ARE BACK IN THE LAND OF FACTORIAL DESIGNS (SEE BOX, ACTIVITY 9).

11.2) Materials

1. microscope slides
2. microscope (one)
3. sticky tape (Sellotape)
4. Petri dishes
5. thin forceps (one–two)
6. razor blades (one–two)
7. opening buffer (50 mM KCl, 0.1 mM CaCl2, 5 mM Mes, pH 6.5) without or with 50 micrometres (µM) ABA

11.3) Procedure

This protocol is based on the ones described in (Roelfsema and Prins 1995; Desikan *et al*. 2002).

Materials	Method
Sellotape, scissors, plant	1. Cut one leaf from each of five plants of between five- and six-weeks old and remove the midvein with a razor blade. CAUTION: place the top side of the leaf (adaxial) down onto a strip of sticky tape.
	2. Cover the other side of each leaf (the bottom or abaxial side, where most of the stomata are) with another strip of sticky tape, pushing gently with your fingertips.
Razor, scissors	3. Trim the sides off before gently removing the strip to recover the leaf peel from the abaxial side. Trim the excess sticky tape with scissors if necessary.
Petri dish, ABA, OB	4. Place the leaf peel attached to the sticky tape strip into a Petri dish containing opening buffer.

Materials	Method
	5. Incubate for 30 minutes under a lamp at 150–200 micromoles (µmol) s^{-1} m^{-2}
Petri dish, ABA, OB	6. Place half of the strips of each genotype into another Petri dish containing the opening buffer plus 50 µM ABA. Leave the other half in opening buffer only.
	7. Incubate for 1.5 hours on the bench (no lamp).
Microscope	8. Begin looking at the control samples (those in the opening buffer) towards the end of the incubation time.
	9. Look at the microscope and record the number of stomata and the number of epidermal cells. Note how many stomata are open, semi open, or closed, and any other feature that you think is relevant. If possible, take photos of one or more microscope field views for each sample to count and analyse stomata. If you want to quantify stomata closure, measure the stomata width and length for a set number of stomata from one leaf on each of three plants per genotype using a calibrated micrometre scale. Express the stomata aperture as the width-to-length ratio.
	10. Examine the samples incubated in the presence of ABA under the microscope after 1.5 hours.
	11. If you take photos of your samples to analyse later, be sure to label them with your group number, date, genotype and plant number. For example: 01–20100705–Col–0–05, is the photo of plant 05, genotype Col–0, taken on 05/07/2010 by group 01.

IMPORTANT: you may want to practice this technique on spare plants before applying to your samples.

During the long incubation, take more leaf peels and observe the anatomy of the epidermal cells under the microscope. This will familiarise you with the epidermal cells (i.e., examine and differentiate guard cells from pavement epidermal cells) when you start to examine the samples in the buffer. This should facilitate the final analyses of the effect of ABA in the stomata closure between genotypes after the two-hour incubation.

11.4) Expected outcome

1. Compare the epidermal peels of the two genotypes. Specifically, look at the epidermal cells and guard cells. Is there the same number of stomata? Are they closed or open?
2. Calculate the following for each photo of the microscope field view:
   a. Stomatal Index; $SI=S/(S+E)$, where S is the total number of stomata, and E the total number of epidermal cells.
   b. Stomatal Density; $SD=S/A$, where S is the total number of stomata present in area A (how would you determine the area?).
   c. Measure the length and width of a few guard cells in each peel, expressed in µm to assess closure.
3. Compare the effect of ABA on stomata closure between wild type and mutant. Are the stomata closed in both genotypes after ABA incubation?
4. Assess whether the differences in stomatal size, number or closure are statistically significant between ABA and control treatments for the wild type and mutant. Remember, this is a factorial design so will require two-way statistical tests (See Appendix B).

Activity 12: Write-up and symposium

Based on your observations and experiments, and in consultation with the web resources provided, try to determine what mutant you were given. In your write-up, you must present evidence to support your claim. Present your data in figures and tables and include drawings as appropriate. You may be able to find an exact match, or you may be able to identify a few likely suspects. You will be marked on the process that you use to narrow down the options, rather than whether you got the right mutant.

All written work must be your own independent work. We encourage students to work together on data entry, analysis and presentation; however, all written work (including the legends of figures and captions of tables) needs to be done individually. Make sure to acknowledge the contribution of the other group members in your acknowledgements section

Follow the format of an original research paper for the journal *Functional Plant Biology* to write your report. Your report will include separate sections for the abstract, introduction, methods, results, discussion and references. In your report:

- Do not exceed 4,000 words (approximately ten pages), not including title page, abstract, tables, figures or references.
- Your manuscript must be double or 1.5-line spaced, single sided, 12-point font with margins of at least 30 millimetres.
- Pages should be numbered consecutively.

IMPORTANT: we will not mark papers that do not follow this format.

12.1) Sections and format of the manuscript

The following instructions were adapted from the 'Notice to authors' contained in *Functional Plant Biology*. The sections of the manuscript should be as follows:

Title page

Use the keywords to devise a concise and informative title. Additional keywords may also be suggested. If you include a botanical name in the title, omit the authority, but include it in the Abstract and at first mention in the 'Methods' section. Please also supply an abridged title, for use as a running head, that does not exceed eight words in length.

Authors and addresses

Please include full first name, initial and surnames for all authors and a current mailing address for each.

Abstract

Scientific papers begin with an abstract, which is a single paragraph without references that summarises the project including what you did, why you did it, and what you found. The abstract should be brief, but informative, without reference to the text. State the scope of the work and the principal findings in fewer than 200 words. References should be listed in full (authors, journal, and volume and page numbers). Scientific names of plants should be accompanied by their authority.

Introduction

The introduction must include background information (including references to published papers) to put your work in context. Aims or objectives of the work, and the significance of the work, should be clearly stated. Include your hypotheses or predictions. Your introduction should not include description of techniques, findings or conclusions.

Methods

The methods section must be concise, but detailed enough to allow your work to be repeated by others. As a guideline, the Methods section should not be longer than 1000 words.

Results

The logic and order of presentation of text, tables and figures is the most important aspect of the results section. Try to present a coherent story that can be followed by your reader. Do not present the same data in both figure and table form. You must include *both* graphical results and a written explanation of the results. TIP: it is a good idea to outline the figures first before writing the results text. See Appendix B for many useful tips on what to do with your data and how to present your results.

Discussion

In your discussion, consider the results in relation to your hypotheses, aims and/or predictions. Place the study in the context of other work (including references to published papers) — and, in doing so, explain any novel contribution that your study might have made.

References

In the text, cite references chronologically; do not number them. Make sure that all references in the text are listed at the end of the paper and vice versa. At the end of the paper, list references in alphabetical order.

Use 'and' to link the names of two co-authors in the text, and use *'et al.'* where there are more than two. Take special care checking the accuracy of the references. ..

Give titles of books and names of journals in full. Include the title of the paper in all journal references, and provide first and last page numbers for all entries and websites.

Managing your references

For Reference Manager, http://www.refman.com/support/rmstyles-terms.asp.

For ProCite, http://www.thomsonreuters.com/endnote.

For EndNote, http://www.crandon.com.au.

You will find the style file under the 'Biosciences' or 'Agriculture' categories, listed as *Functional Plant Biology.*

Figures

Refer to each figure or illustration in the text. Legends to figure axes should state the quantity being measured and be followed by the appropriate SI units in parentheses.

Tables

Refer to each table in the text. Number each with an Arabic numeral and supply a heading. Use **Table Formatting** (i.e., use table cells); do not use tabs, spaces or hard returns when setting up columns.

Avoid long titles by incorporating an explanatory note below the title, which should be started on a new line from the title of the table. Include in the head note, where applicable, the levels of probability attached to statistics in the body of the table, and any other information relevant to the table as a whole.

Avoid footnotes where possible; use them only to refer to **specific and single** data points in the body of the table. Use A, B, etc for footnotes; use *, **, *** only to define probability levels. Insert horizontal rules above and below the column headings and across the bottom of the table; **do not use vertical rules**. The first letter only of headings of rows and columns should be capitalised. Include the symbols for the units of measurement in parentheses below the column heading. Use standard SI prefixes with units in the column headings to avoid an excessive number of digits in the body of the table.

Accepted abbreviations

The following terms are accepted abbreviations, and do not require full explanation in text: ABA, ADP, ADPase, ANOVA, ATP, ATPase, bis-Tris, DMSO, DNA, EDTA, GA, Hepes, HPLC, IAA, MES, Mops, NAD or NAD^+, NADH, NADP or $NADP^+$, NADPH, $NADP^+$, PAGE, PBS, Pipes, PSI, PSII, RNA, Rubisco, SDS, Tris, UDP, UV.

Units

Use the SI system where appropriate, especially for exact measurement of physical quantities. However, non-SI units such as day and year are acceptable. Give measurements of radiation as irradiance or photon flux density, or both, and specify the waveband of the radiation. Photon flux density units should be used in papers concerned with the quantum efficiency of plant photo processes. Do not use luminous flux density units (e.g., lux). Use the negative index system; e.g., $g\ m^{-2}$, $kg\ ha^{-1}$, $mm^{-2}\ s^{-1}$.

Statistical evaluation of results

Describe the experimental design and analysis in sufficient detail to allow them to be evaluated, and support this description with references if necessary. State the number of individuals, mean values and measures of variability. State clearly whether you have included the standard deviation or the standard error of the mean.

Mathematical formulae

Adequately space all symbols. Avoid two-line mathematical expressions in the running text. Display each long formula on a separate line with at least two lines of space above and below it.

Acknowledgements (optional)

This section presents an opportunity to express thanks for the technical assistance or other support that you have received. Acknowledge students with whom you collaborated on data analysis.

References

For formatting instructions, see Notice to Authors' at http://publish.csiro.au/nid/105/aid/413.htm#4. Except under unusual circumstances, the list of literature cited will not include web references, but only references to peer-reviewed published articles or books. Use Endnote to create the list of literature cited.

12.2) A few additional suggestions for improving scientific writing

- Species and genus names should always be italicised (or underlined). If you have already given the full species name, you may abbreviate the genus name to the first initial. Do not abbreviate the species name. Do not abbreviate the genus name at the beginning of a sentence. Do not abbreviate the genus name if there is likely to be any confusion (e.g., if your study organisms have the same species names but different genus names, or if you study several different species with genus names all beginning with the same letter).

   Yes: *Eucalyptus benthami*, *E. benthami*,

   No: Euc. ben., Eucalyptus benthami, E. benthami

- Be careful with the use of the terms variance and significant in scientific writing. These words have specific meanings in statistics, and should only be used in their statistical sense. Don't use 'significant' if you mean 'a large difference', unless you can provide a statistical result to support the claim.

- In an experiment like ours, it is not accurate to say 'root mass decreases at low light'. It is correct to say 'root mass was lower at low light'. To say that root mass has decreased implies that root mass at harvest was lower than it was a planting.

- The following phrases are not necessary and should not be used in scientific writing: 'It is known that', 'It was shown that', 'It can be seen that'. Read the sentences below. You will see that these phrases do not change or add to the meaning of the sentences. Scientific writing should be concise — any words that do not add to the meaning of the sentence are left out.

   Yes: 'The species responded differently to the light levels'

   No: 'It was shown that the species responded differently to the light levels'

   Yes: 'Plants grow faster at high light'

   No: 'It is known that plants grow faster at high light'

   Yes: 'Wild type plants responded to the light treatment to a greater extent than the other species'

   No: 'It can be seen that wild type plants responded to the light treatment to a greater extent than the other species'

- When referring to figures and tables in a scientific write-up, do not write 'Fig. 1 shows that ...'. Instead, describe the result and refer to the figure or table in parentheses at the end of the sentence. This approach is more concise, and draws the reader's attention directly to the result you are describing.

   Yes: 'SLA was higher in mutant plants grown under drought (Fig. 1).'

   No: 'Fig. 1 shows that SLA was higher in mutant plants grown under drought.'

   No: 'Fig. 1 shows the result of the comparison of SLA. SLA was higher in mutant plants grown under drought.'

- <u>References</u>: all sources listed in the reference section should be cited in the text of the paper. Likewise, all sources cited in the text should be listed in the reference section. Be careful to use the correct citation format. In the case of your write-up, apply the format followed by the *Functional Plant Biology*.

12.3) Symposium

In Week 12 we will meet during the laboratory session to present and discuss your group's plant detective outcomes. Each group should prepare a 10–15 minute presentation on their results. We recommend you use power point. Divide the responsibility for preparing and delivering between your group.

In your presentation summarise the results of your investigations, highlighting how your mutant differed from the wild type. Tell us what mutant you have concluded it is, and why you reached that conclusion. Describe how the mutation has affected both form and function and what its consequences would be under natural conditions.

The symposium presentation will receive a group mark (5%) and an individual mark (5%).

Appendices

Appendix A: General rules for safety and conduct

Safety in the laboratory is both an individual and management responsibility. As an individual you must observe the following general rules which have been extracted from Australian Standard 2243, Part 1 — 1982, 'Safety in laboratories Part 1 — general' (1982:6), published by the Standards Association of Australia. In addition, all students must wear protective clothing, appropriate footwear (not thongs), and safety glasses where appropriate.

1. Never adopt a casual attitude in the laboratory.
2. Never run in the laboratory or along corridors and do not indulge in horseplay.
3. Minimise clutter on the workbenches and the floor beneath the benches. Please place your personal gear (e.g., bags and jackets) on the benches at the back of the laboratory.
4. Ensure that clothing is suitable for the laboratory:
    - wear covered, non-slip footwear — no thongs, sandals or scuffs
    - do not wear loose clothing, particularly scarves
    - wear a laboratory coat for the duration of the class during any laboratories where you may be affected by chemical spills
    - ensure that long hair is tied back
    - do not wear earphones of any sort
    - gloves are worn to protect your skin from contact with hazardous chemicals. Dispose of gloves immediately after they come in contact with chemicals as the glove matrix will break down. Also remove gloves before touching clean objects, such as pens, benches or door handles to prevent the spread of hazardous substances
    - contact lenses should not be used when working in the laboratory as they are known to absorb chemicals and concentrate them on the surface of the eyes. We recommend eye glasses are worn instead
    - safety glasses **must** be worn at all times.
5. Do not handle or consume food or drink in laboratories.
6. Mobile phones should be turned off before the class and not handled until after the class (after hands have been washed).
7. Never undertake any work unless you are aware of the potential hazards of the operation. Take note of these hazards and follow appropriate safety precautions. Regard all substances as hazardous unless there is definite information to the contrary. Read the Material Safety Data Sheets (MSDS) for full information.
8. Do not work alone in a laboratory; a second person should always be within call.
9. Please notify a member of staff immediately of breakages or spillages.
10. In case of a spill get help immediately. Wash contaminated skin with water for 20 minutes, do not use soap or solvents to remove chemicals (unless advised to do so). Know where the eyewash and emergency showers are located.
11. Always use safety carriers for transporting glass or plastic containers with a capacity of 2 litres or greater. Exercise particular caution when carrying containers of mutually reactive substances.
12. **Always wash your hands thoroughly after any work in the laboratory**.
13. Exercise care when opening and closing doors and entering or leaving the laboratory.

Appendix B: What do I do with my data?

B.1) An introduction to data

Data are wonderful (and always plural). In the Plant Detectives Project you will produce a surprising amount of data and we want to get you off on the right foot so you know how to record and organise that data. This may seem basic, and even unimportant, but just imagine you're listening to years of previous students saying 'I really wish I'd paid more attention to how I collected my data in the first weeks of the project', and you might decide to follow these tips from the start!

Ongoing and efficient data collection and record keeping puts you in a position to analyse your data and ask questions that you may not be able to anticipate at the outset. So, try to keep all bases covered. We will demonstrate these techniques in principle, but not with a specific software package in mind. Your instructor will provide you with instructions tailored to the packages you have available.

PLANT DETECTIVE TIP

FOR PLANT LABELS:
LABEL EACH PLANT ACCORDING TO THE FOLLOWING FIELDS:
1. GROUP
2. PLANT
3. GENOTYPE: M, MUTANT; OR, WT, WILD TYPE
4. CONDITION: FOR WHEN WE APPLY AN EXPERIMENTAL TREATMENT (E.G., WW, WELL WATERED OR D, WATER STRESSED)
EXAMPLE: THE LABEL '4.1 M D' IS FOR GROUP 4, PLANT 1, WHICH IS A MUTANT THAT WILL BE SUBJECTED TO WATER STRESS OR DROUGHT.

FOR TUBE LABELS:
WHENEVER YOU PUT SAMPLES IN TUBES OR DISHES FOLLOW THE SAME LABELLING CONVENTION.

1. Labels

a. As described in the practical activities and appendices, label every plant with a unique number, indicate whether it is wild type or mutant, and provide the background genotype you have. Make sure the labels stay with the plants. To avoid losing a label, consider painting the label onto the pot containing the plant.

b. Whenever you record data, make sure you also record the information on the label.

2. Datasets

The data that you collect should first be entered into a spreadsheet or database. Your instructor will advise on what package to use, but the following rules apply to most:

a. Put the column labels in the first row — do not put other text in the rows. Use comments or a notes page for additional text

b. Include columns for the following in all of your datasets:

    i. plant ID: this is the plant number, if you decide to serial number your samples, or the pot number

    ii. genotype: wild type or mutant

    iii. replicate: number of the replicate

    iv. date: the date of data collection. You may also want to include the practical week, since this will be used when graphing the data

    v. name of collector: this is optional. If you record who in your group collected the samples it can be useful in the future if something goes missing or looks confusing

    vi. treatment (if applicable)

c. Always record the units of your measurements; it is surprisingly easy to mix up centimetres (cm) and millimetres (mm) after a period of time has passed without reviewing data

d. Do not leave blank lines between your rows or columns. Blank lines confuse the sorting algorithms of your spreadsheet program and will lead to the sorting of some, but not all, data. At the least this is annoying. At the worst you end up sorting your labels and not your data or vice versa.

e. Save and back up your data regularly:

    i. take turns within your practical group to upload data to an online storage option each week

    ii. get agreement from the group on the dataset to be used and to keep it backed up.

Sample data sheets

A. Record of major phenotypic observations

Experiment:

_____

Date start:

_____

Seeds

Date	Day	Comment/observation

B. Excel spreadsheet to record data — always record units too

Student or group name	Plant number	Genotype	Treatment (if applicable)	Date and/or week in practical	Trait value 1 (e.g., stem height)	Trait value 2 (e.g., leaf size

C. Alternative spreadsheet format, for germination assay

Plate #	Position #	Genotype	Week	Date	Trait value 1 (e.g., root length (mm))	Trait value 2 (e.g., secondary roots (1/0))

B.2) Exploratory data analysis

In Section 1 we discussed how to record and organise your data. This section covers how to explore your data and present it visually. This is called exploratory data analysis (EDA). While these techniques can be demonstrated in principle, this section has not been structured with a specific software package in mind. Your instructor will provide you with guidelines tailored to the packages you have available.

Why is this important? Too often researchers (not just students) leap into analysing their data statistically before getting to know what it's telling them. EDA is an important first step that can save time in the long run.

B.2.1) Look for outliers and check the distribution of your data points

The following analyses are based on the assumption that your data will fit, roughly, a normal (bell-shaped) distribution. But, data don't always oblige. You need to check the distribution of your data before proceeding.

Further, you need to check for outlying points — outliers — in your data. Most frequently these are data entry errors, but sometimes you will identify a data point that is different from the rest. This point may cause your data to violate the assumptions of normality. You may need to exclude the point or to transform your data to make it approximately normal.

You can use a statistical analysis package to plot a histogram or box plot of your data to check for normality and look for outliers:

a. If your data are normally distributed, your histogram will look like a bell curve. If your bell has long tails or an isolated point shows up on one or other side of the curve, you have an outlier. If your data does not form a pretty single hump or your curve leans heavily in one direction or the other, you may have data that are not normally distributed. You may need to use a mathematical transformation to rescale your data so they meet the assumption of normality. The most common transformation in biology is the log transformation. Don't fear transforming your data: you have not changed them, just scaled them.

b. A box plot is an alternative way of looking at your data. A box plot (Fig. i) shows you the mean (average), median (middle value), and the quantiles of your data. The 25% and 75% quantiles are generally shown as the top and bottom of your 'box'. The median (50% quantile) is shown as a line drawn across the box. The mean is generally a dot or diamond that should sit near to the median. The box usually has whiskers to show you where the lowest 5% and highest 95% of your data sit.

Any points that appear above or below the whiskers are outliers. If your mean is skewed to one side of your box, your data are not normally distributed.

You can also use a spreadsheet program with chart options to look for outliers by making an x v y plot, or scatter plot, of two columns of your data. For example, leaf number vs rosette diameter, where you have both measures for each plant. In this case, it is likely that these two variables would scale closely with each other. Any plant that has a high leaf number for its rosette diameter, or a high diameter for its leaf number will fall at a distance from the other points on your graph. Check these points to see if there has been a data entry error (most often this is a mistakenly added or removed digit or decimal place. If most plants with a rosette diameter of 5 cm have ten leaves, but one of your plants has one or 100 — it's probably a typo.)

What do you do with an outlier?

a. First you must decide whether the outlier is biologically real or a data entry error. If the latter, fix the mistake. If possible, repeat the measurement. If not possible, then you may choose to delete the entry from your dataset and make a note in a comments column explaining why a data point was deleted. These are called missing data and should be mentioned in your write-up.

b. If the point is not a typo but is a statistical outlier you may still need to exclude it. Determine whether the value reflects a plant that has somehow been damaged or is ill or whether you have had an instrument failure that has given you an incorrect result. If so, delete as above.

c. Finally, sometimes you will come across an individual that is just very different from the others. In the case of this practical, this might even be a seed that has mistakenly been mixed into a seed lot, or a spontaneous mutation. Such data points, even though real, will prevent you from being able to statistically analyse your data for two reasons. First, the points may lead you to violate the assumption of normality. Second the data points will have undue influence on your ability to estimate the average value of the variable you are measuring. For example, if you have one plant out of 24 that is ten times the size of the others, you would calculate an average size that is much larger than 23 of your 24 plants. This is not an accurate description of the average size. In such cases you are justified in excluding the data point from your graphs and statistical analysis. You must, however, mention in your write-up that the data were excluded and why.

d. Note: if you delete a data point because it is an outlier, make sure to do so in all copies of your dataset and keep records — that way you won't get confused down the track.

B.2.2) Compare the average (mean) of your two populations (wild type and mutant)

Once you are sure that you have entered your data correctly and there are no outliers, it is possible to compare the mean of the two populations. By 'population' here we mean wild type versus mutant. Or, if you have imposed a treatment like a drought stress, you will have two populations * two treatments = four means to compare. To compare the populations, calculate the mean and the variance. The variance describes how much difference there is among the plants within each of your populations. If the variance is large, then it will be more difficult to see a statistically significant difference in the population means.

Most spreadsheet packages will enable you to calculate a mean and standard deviation (a common descriptive measure of variability in your data) using formula or equation functions. If you wish to calculate these yourself, the mean is simply the average (sum the values and divide by the total number of points). The standard deviation (SD) is calculated as:

$$SD = \sqrt{Variance} = \sqrt{\frac{\Sigma(X - \bar{x})^2}{(n-1)}}$$

Where X is the value for a given sample, $\bar{x}$ is the mean of all samples, and n is the total number of samples of a given treatment or genotype combination (e.g., number of wild type plants measured).

When you want to compare averages (means), for example, across treatments or genotypes, a measure of the variability around the mean itself is required. This measure of variability is the standard error, which is a measure of the variability in the mean itself and is used to infer whether two sampled means are different. To calculate standard error, use SD and n:

$$SE = SD/\sqrt{n}$$

B.2.3) A picture is worth 1,000 words

Plot a graph of your data so that you can visualise the differences (or lack thereof) between the average outcome of your wild type and mutant plants. A simple bar chart will usually suffice. Be sure to plot error bars on your graph so that you can see the variance around your mean. If the error bars of your means are heavily overlapping, don't expect a statistical test to tell you there is a difference in the means!

We recommend that you graph your data at the end of each week's practical. This way you keep up to date on what your results are. You should bring your graphs to the start of the next practical class so that you can present your results in your discussion group and find out if the other student groups got the same results. If your results differ, you may have found something that distinguishes your mutant.

As most spreadsheet packages have graphing options that are adequate to make the graphs needed for this project, and some statistical software packages also make excellent graphs, it is not necessary to use a dedicated graphing package. And, if you are so inclined, graph paper and rulers still provide a perfectly effective way of graphing your results!

B.3) Comparing two means

T-tests (or one-way analysis of variance, ANOVA) are used to assess whether the average value for a trait in one population differs from that of another. Specifically these tests assess whether the between population variation (e.g., between the means) is greater than that within the populations. In this case the populations are the wild type and mutant plants and the trait of interest may be rosette diameter or leaf size. The T-test is the statistical formalisation of the graphing of data that revealed overlap in the error bars.

All statistical packages and most spreadsheet packages have an option for calculating a T-test. The formatting requirements vary among packages. If the package requires setting a hypothesised mean difference, set this to 0. You may also need to set an alpha level; pick 0.05 (see below).

To run the T-test, supply the following information:

- the measured variable that you are interested in comparing, e.g., root length or rosette diameter, which is known as the data variate, **response variable**, or **dependent variable**; statistical packages differ in their choice of language

- the variable indicating group or population, which is known as the group **factor** or **independent variable**. In this case the population is genotype: wild type or mutant

- '**discrete**' and '**continuous**' variables. Not surprisingly, continuous variables are those for which the values vary continuously. A factor or variable is called discrete when it can only have particular states — low, medium and high, for example. For a T-test, the population variable is discrete and the measured variable is continuous

- **degrees of freedom**, which reflect the number of data points, or samples, in your analysis. For a T-test the degrees of freedom is the number of samples in each population minus 1; or, the total number of samples minus 2

- the *P* value, which tells you the probability that the observed difference in means could have arisen by chance. Small values of P indicate that this difference was unlikely to have arisen by chance, and we call this evidence for a true difference. By convention *P* values are generally defined as less than 0.05 to be sufficiently small to count as evidence for a true difference. This a 'statistically significant difference'.

When you run your T-test, specify the measured variable and the group variable and get the package to calculate the test statistics. Your table will look something like this:

```
Summary

                sample              Standard    Standard
                        Mean  Variance        error of
                 size              deviation    mean

WT                 84   11.29   17.97    4.239     0.4625

mutant             56   11.5    13.78    3.712     0.4961


Difference of means:              -0.214
Standard error of difference:      0.696

95% confidence interval for difference in means: (-1.591, 1.163)


Test of null hypothesis that mean of Nodules with Hydroxylation_P2 = 0 is equal to
mean with Hydroxylation_P2 = 1

Test statistic t = -0.31 on 138 d.f.
```

In this case, the most important things to be able to glean from this table are the means, the standard errors of the means, the group size and the P value. The degrees of freedom should be two less than your number of samples. (If P < 0.05 then it is legitimate to refer to a statistically significant difference in the mean value between your genotypes).

Think about your results, compare that statistical result to your observations of the means from your exploratory data analysis — do the statistics support your previous conclusions?

B.4) Two-way tests

Up to this point we have discussed statistical tests comparing two populations only: wild type and mutant. For some studies that you will do in the Plant Detectives Project, however, you will consider two experimental factors at once; these are factorial designs (See box in Activity 9). For example, in the drought experiment you will consider the effect of water stress as well as genotype, asking: Do the mutants differ from the wild type plants in their response to water stress? If your drought-stressed mutant plants fare better than your drought-stressed wild type plants, then you will conclude that the mutation improves the ability of the plants to cope with drought. Sometimes the effect of a mutation will not be apparent under benign conditions: only under stressed ones. In these cases we say there is an interaction between the genotype and drought effects if the severity of the drought effect depends on genotype.

To run a two-way ANOVA, you need to supply similar information to that gathered for the t-test:

- the measured variable that you are interested in comparing; e.g., root length or rosette diameter
- the variables indicating the populations/treatments. In your case the population is genotype: wild type or mutant, and the treatment is normal or drought
- indicate that you want to compare means by genotype and by drought treatment, as well as a comparison of the drought response by genotype. In the treatment structure box, write:

genotype + drought + genotype.drought

A two-way ANOVA produces a table like this:

```
Analysis of variance

Variate: leaf_area

Source of variation     d.f.        s.s.        m.s.      v.r.    F pr.
genotype                  1         426.5       426.5     2.39    0.126
drought                   1        7001.6      7001.6    39.29    <.001
genotype.drought          1         600.3       600.3     3.37    0.070
Residual                 74       13185.6       178.2
Total   77      21213.9


Probability = 0.759
```

From this table you want to assess three different P values. The P value for the genotype effect tells you whether or not the genotype means (ignoring drought) differ. The P value for the drought effect tells you whether or not the drought treatment means (ignoring genotype) differ. The interaction term P value tells you whether the effect of drought on wild type is the same as the effect on the mutant.

In our case, interaction term P value is the one of interest, and small P values for the interaction term provide evidence that the effect of drought on mutants is different to the effect of drought on wild type plants. The interaction term may be significant if one type responds more than the other species (for example if the wild type is less sensitive to drought). Alternatively, the interaction term may be significant if one type responded in the opposite direction to the other, or didn't respond at all. If you have significant type and treatment effect, but don't have a significant interaction term, the analysis indicates that each type responds to light in the same way, but that the types differ in their inherent root mass ratios.

Again, your stats package should also calculate the means and standard error for your treatments. There will be three sets of means: the genotype means and drought means (each of which ignores the other treatment) and the means of all four groups (wild type under drought, wild type well watered, mutant under drought and mutant well watered).

What do you do with this information? The test statistics can be reported in a table or in text and the means can be graphed to illustrate significant results. If there is not a significant P value for the interaction of genotype and drought treatment,plot the means of each, ignoring the other. Otherwise, plot the four means in a bar or line graph. Sometimes the lines make it easier to interpret the results, sometimes bar charts are more effective. It is up to you which you present (see Fig. 12).

Two-way ANOVA table

Source	A	B	C	D
Genotype	≤0.05	ns	≤0.05	≤0.05
Drought treatment	ns	≤0.05	≤0.05	≤0.05
Genotype * Drought treatment	ns	ns	ns	≤0.05

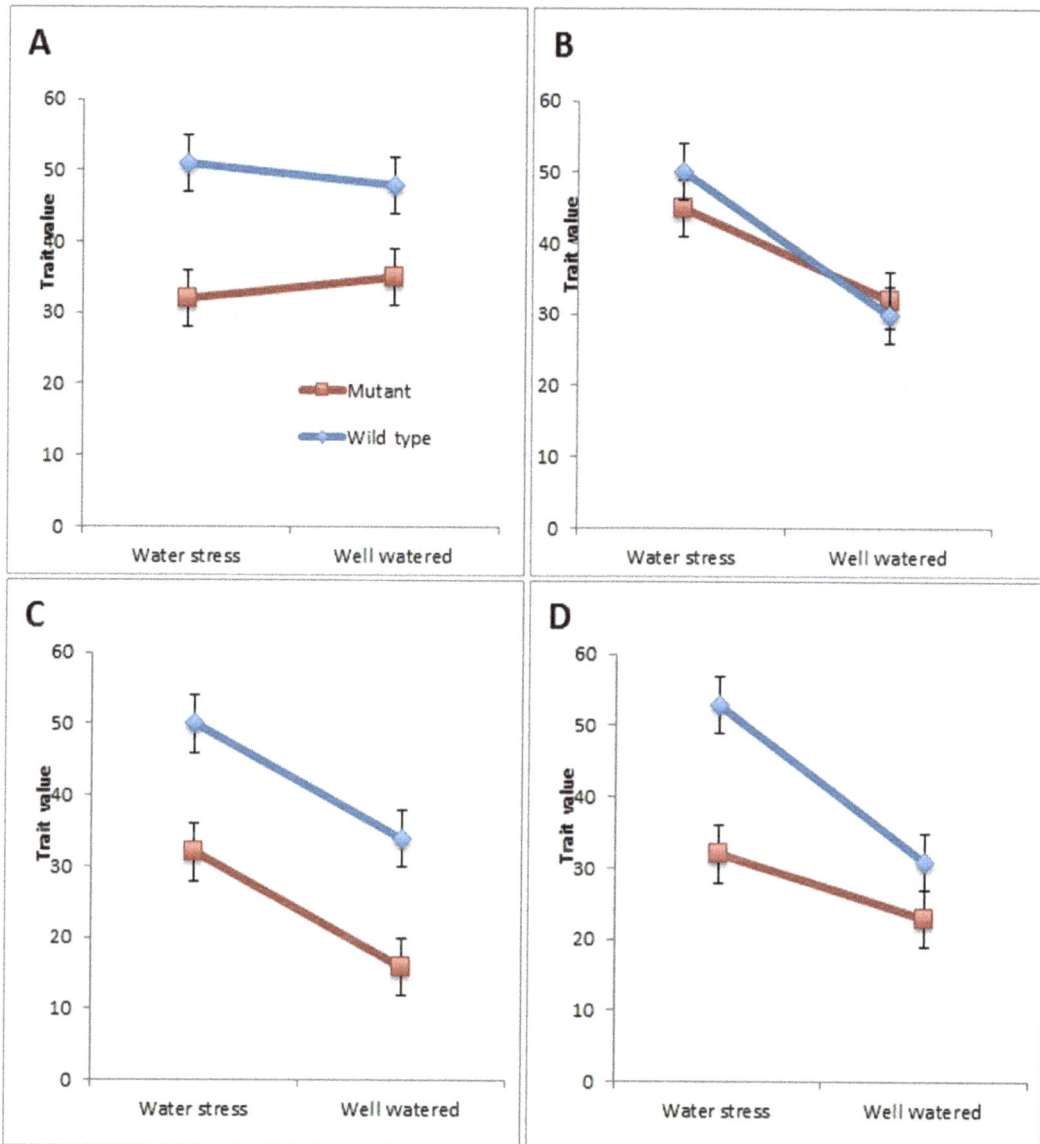

Figure 12: P values for several hypothetical two-way analyses of variance are given in the table
The scenarios A, B, C and D are graphed in the interaction plots of the same label. Note particularly the slopes of the lines and the overlap (or lack thereof) of the error bars.

B.5) Presenting your results

For your final report include tables and figures that present your results as described in Activity 12.1. There is an art to presenting your data well — we do not want to see your raw data, nor do we want exhaustive tables of mean values or pages and pages of statistical analyses. Rather, we want you to use your data to tell a story — in this case, to make an identification of your mutant Arabidopsis and to justify that identification. As the semester progresses, you will accumulate a substantial dataset and a wide range of results from your work. You may choose to present all or just some of the data you have collected and analysed — sometimes less is more if it helps get your story across more clearly.

Remember the following when presenting your results:

B.5.1) Tables

a. tables have a brief descriptive title **above** them and columns below the title
b. the first line of the table should contain column identifiers
c. ensure that all values show units (e.g., cm or grams (g)). Tables can be made in word processing or spreadsheet programs
d. do not present the same data in both a table and a graph

B.5.2) Figures

a. figures have a brief descriptive legend **below** them. Either describe in the legend or show on the figure the meaning of symbols and abbreviations
b. label all axes and show the units (e.g., cm or g).
c. include error bars with SD (or standard error if that is what your statistics package calculated) on them. Specify what the error bars are in your legend
d. figures can be made in a range of software packages or they can be drawn by hand using graph paper. Do whatever you prefer, but make sure the points above are adhered to for all graphs

B.6) Statistical description of two-sample tests

For those of you interested in the guts of the T-test: the equation we use for this test is:

$$t = \frac{\bar{X}_1 - \bar{X}_2}{s_{\bar{X}_1 - \bar{X}_2}}$$

Where

$\bar{X}_1$ and $\bar{X}_2$ are the means of samples 1 and 2 respectively. And, where n_1 and n_2 are the sample sizes (number of data) for samples 1 and 2 respectively. And,

$$s_{\bar{X}_1 - \bar{X}_2} = \sqrt{(s_p^2/n_1) + (s_p^2/n_2)}$$

Where DF1 and DF2 are degrees of freedom $(n - 1)$ and

$$SS = \sum (X - \bar{X})^2$$

for samples 1 and 2 respectively.

When you have calculated your t-statistic, look up the t value in this table: http://www.itl.nist.gov/div898/handbook/eda/section3/eda3672.htm. The following information is based on (Bower *et al.* 1989).

Appendix C: Arabidopsis growth stages

Stage	Description	Col-0 Data		
		Days[a]	sd	CV[b]
Principal growth stage 1	Leaf development			
1.02	2 rosette leaves >1 mm in length	12.5	1.3	10.7
1.03	3 rosette leaves >1 mm in length	15.9	1.5	9.5
1.04	4 rosette leaves >1 mm in length	16.5	1.6	9.8
1.05	5 rosette leaves >1 mm in length	17.7	1.8	10.2
1.06	6 rosette leaves >1 mm in length	18.4	1.8	9.8
1.07	7 rosette leaves >1 mm in length	19.4	2.2	11.1
1.08	8 rosette leaves >1 mm in length	20	2.2	11.2
1.09	9 rosette leaves >1 mm in length	21.1	2.3	10.8
1.1	10 rosette leaves >1 mm in length	21.6	2.3	10.9
1.11	11 rosette leaves >1 mm in length	22.2	2.5	11.2
1.12	12 rosette leaves >1 mm in length	23.3	2.6	11.3
1.13	13 rosette leaves >1 mm in length	24.8	3.2	12.8
1.14	14 rosette leaves >1 mm in length	25.5	2.6	10.2
Principal growth stage 3	Rosette growth			
3.2	Rosette is 20% of final size	18.9	3	16
3.5	Rosette is 50% of final size	24	4.1	17
3.7	Rosette is 70% of final size	27.4	4.1	15
3.9	Rosette growth complete	29.3	3.5	12
Principal growth stage 5	Inflorescence emergence			
5.1	First flower buds visible	26	3.5	13.3
Principal growth stage 6	Flower production			
6	First flower open	31.8	3.6	13.3
6.1	10% of flowers to be produced have opened	35.9	4.9	13.6
6.3	30% of flowers to be produced have opened	40.1	4.9	12.3
6.5	50% of flowers to be produced have opened	43.5	4.9	11.2
6.9	Flowering complete	49.4	5.8	11.7

Stage	Description	Col-0 Data Days[a]	sd	CV[b]
Principal growth stage 1	Leaf development			
1.02	2 rosette leaves >1 mm in length	12.5	1.3	10.7
1.03	3 rosette leaves >1 mm in length	15.9	1.5	9.5
1.04	4 rosette leaves >1 mm in length	16.5	1.6	9.8
1.05	5 rosette leaves >1 mm in length	17.7	1.8	10.2
1.06	6 rosette leaves >1 mm in length	18.4	1.8	9.8
1.07	7 rosette leaves >1 mm in length	19.4	2.2	11.1
1.08	8 rosette leaves >1 mm in length	20	2.2	11.2
1.09	9 rosette leaves >1 mm in length	21.1	2.3	10.8
1.1	10 rosette leaves >1 mm in length	21.6	2.3	10.9
1.11	11 rosette leaves >1 mm in length	22.2	2.5	11.2
1.12	12 rosette leaves >1 mm in length	23.3	2.6	11.3
1.13	13 rosette leaves >1 mm in length	24.8	3.2	12.8
1.14	14 rosette leaves >1 mm in length	25.5	2.6	10.2
Principal growth stage 3	Rosette growth			
3.2	Rosette is 20% of final size	18.9	3	16
3.5	Rosette is 50% of final size	24	4.1	17
3.7	Rosette is 70% of final size	27.4	4.1	15
3.9	Rosette growth complete	29.3	3.5	12
Principal growth stage 5	Inflorescence emergence			
5.1	First flower buds visible	26	3.5	13.3
Principal growth stage 8	Silique ripening			
8	First silique shattered	48	4.5	9.3
Principal growth stage 9	Senescence			
9.7	Senescence complete; ready for seed harvest	NDc	ND	ND

[a] Average day from date of sowing, including a three-day stratification at 4 °C to synchronise germination

[b] CV, coefficient of variation, calculated as (SD/days) x 100

[c] ND, not determined (see text for details)

Adapted from Table 2 of (Boyes *et al.* 2001): Arabidopsis growth stages for the soil-based phenotypic analysis platform. Visit http://www.arabidopsis.org/portals/education/growth.jsp to see the original table.

Note that stages described and timeline are for plants grown under the following conditions:

Columbia ecotype plants were grown in soil in 16-hour days/eight-hour nights with temperatures of 21°C during the day and 21°C at night. Lighting was provided with fluorescent bulbs giving an average light intensity of 175 micromoles (µmol) meter^{-1} second^{-2}. Seeds were cold treated (stratification) for three days at 4 °C after imbibition to synchronise germination. Days until each stage are approximate and will vary according to growth conditions and genetic background. Approximate dates given include the three-day stratification.

Appendix D: Databases and web resources

Links mentioned in the text

The table below lists the URL of important links organised by activity (A) number. The Content and URL of the link are also indicated.

A	Content and URL
	The Arabidopsis Information Resource (TAIR): www.arabidopsis.org
	The European Arabidopsis Stock Centre (NASC): www.arabidopsis.info/
1	Plant Phenomics Teacher Resource — Plant growth and analysis activity: www.plantphenomics.org.au/education/resources/PlantPhenomics_TeacherResourceBooklet_2013_PRINT-proof_130627.pdf
2	Gene Networks in Seed Development: seedgenenetwork.net
2	The Seed Biology Place: www.seedbiology.de/germination.asp
2	*Arabidopsis thaliana*: A model for the study of root and shoot gravitropism: www.bioone.org/doi/full/10.1199/tab.0043
2	Seed sterilisation and plating: www.youtube.com/watch?v=MdnHvzON5ak&feature=youtube_gdata_player
7	Using the LI-6400 Manual ftp.licor.com/perm/env/LI-6400/Manual/Using_the_LI-6400XT-v6.2.pdf
7	Gas exchange measurements of photosynthetic response curves: prometheuswiki.publish.csiro.au/tiki-index.php?page=Gas+exchange+measurements+of+photosynthetic+response+curves
1 2	*Functional Plant Biology* – 'Notice to authors': www.publish.csiro.au/nid/105.htm

The Arabidopsis Information Resource (TAIR)

TAIR is one of the most comprehensive Arabidopsis seed databases. Check if your institution is subscribed to the database; alternatively access NASC.

There are several ways to access the information on TAIR site:

1. Start at the Germplasm Search
   (http://www.arabidopsis.org/servlets/Search?action=new_search&type=germplasm): Enter a phenotype of interest in the search box next to the words 'phenotype/description (e.g. round leaves)' and click on Submit Query. Alternatively, request 'has images' in the Restrict by Features section.

2. Start at the ABRC Catalogue Browser (http://www.arabidopsis.org/servlets/Order?state=catalog): Seed Stocks: I. Mutants: Characterized Lines. Those with camera icons beside the entries have pictures which you can consult.

3. For a text file with all the phenotype descriptions linked to a locus, visit the TAIR FTP site: ftp://ftp.arabidopsis.org/home/tair/User_Requests/OLD/Locus_Germplasm_Phenotype.01292008. Check if your Institution is subscribed to TAIR. You may have access to an updated list.

4. Collection of databases: http://csmbio.csm.jmu.edu/biology/courses/bio455_555/atlab/database.html

Appendix E: Software tips

E.1) Scanalyzer software output and data interpretation

The following concepts and definitions are adapted from the LemnaLauncher and LemnaMiner Manual (v. 10.03.08) from Lemnatec.

Area

Area is the number of pixels there are in an object. Note that areas with different colours can be estimated. These colours relate to differences in pigmentation produced by genotypic difference, developmental stage or growing conditions. Most quantitative measurements of size are based on area measurements. Based on these values and colour classification, objects can be characterised to identify colour distributions, or area can be related to biomass or movement of organisms.

Compactness

Compactness is defined as the ratio between the area of the object ('ObjectArea') and the area of the ConvexHullArea (the smallest geometrical object without concave parts that covers the whole object). In simpler terms, compactness provides an important quantitative value describing the subjective visual assessment of being compact.

MinEnclosing Circle diameter

This value describes the radius of the minimum hole the object would be able to pass through. This radius refers to the manually less stringently definable maximum 'diameter' that is often measured with a caliper. It is particularly suitable for asymmetrical objects that differ greatly from circles.

Eccentricity

As symmetry and regular growth are common in nature, the extent of non-centric shape can describe important effects; e.g., deviations from the regular growth scheme. Values range from 0 (corresponding to a circle) to 1 for a line.

$$\text{Eccentricity} = \text{sqr}(\mu 20 - \mu 02) + 4 * \text{sqr}(\mu 11)/\text{sqr}(\mu 20 + \mu 02)$$

Below is a grid representation showing the columns and row of a 24 cell grid. Note that the red tape on the top right of your tray should help you to identify what genotype is in each cell.

Red tape reference

1 2 3 4 5 6

A
B
C
D

E.2) Photoshop

This brief 'how to' for optimising images in Photoshop is useful for editing scanned images, such as of MS agar plates. There are three steps in optimising your root scan: auto levels, manual adjustment and sharpening.

Auto levels: Selecting this will automatically adjust the black and white point of your image, making black parts darker and white points lighter:

1. select the roots and the area around them, avoid the white label
2. go to Image: Adjustments: Auto levels

   OR, use the keyboard shortcut: CTRL+SHIFT+L. Your roots should suddenly look a lot clearer.

Manual adjustment: This is a good way to make your roots clearer, and a backup option in case there is limited improvement by selecting auto levels:

1. select the roots and the area around them, avoid the white label
2. go to Image: Adjustments: Levels
3. Select OK when the roots become most clear

Sharpening: If there are fuzzy edges on your scan, sharpening the image can make the roots easier to see. This step is optional, and usually unnecessary:

1. select the entire image (CTRL+A)
2. Go to Filter: Sharpen: Sharpen

You can repeat this if the first sharpening doesn't work, but keep in mind that the more you sharpen your image, the more pixelated it will appear.

E.2) ImageJ

ImageJ, a useful image-processing application that is freely available for download, can be used for root length analysis. Visit http://imagej.nih.gov/ij to download ImageJ.

There is an online manual available from the same link and an abundance of user-generated 'how-to' forums can be found by 'googling' your query. It is an excellent tool for taking post-hoc measurements, such as root length or leaf area, and is simple to use. The following instructions on how to measure root length can also be used for taking leaf area or other relevant measurements.

1. Begin by taking an image. The best images of plants grown on plates are obtained using a flatbed scanner with the agar face (not the lid) of the sealed plate facing down on the glass. Include a ruler as a scale and use a piece of black cardboard for the background. ImageJ can be used with most image formats (e.g., jpeg or tiff).

Open ImageJ. To open an image file, click File: Open. Select your image.

Set the scale.

    a. Use the line tool to draw a 1-centimetre (cm) line, using the ruler in your image as a guide. Press and hold the 'shift' key for a straight line.

    TIP: zoom by using the zoom tool on the tool bar, or simply hover the mouse over the area you wish to zoom on and press the '+' or '−' key.

    b. Click on Analyse: Set scale. In the dialogue box, enter 1 in the known distance box and change the unit of length to cm. Tick the global box — this applies the scale to all of the images you open in this session. If you have different scales for different images, set the scale first! Notice that the distance in pixels box will automatically change based on the known distance.

2. Now you are ready to trace and measure the roots.

    a. Right click the line tool icon and change the setting to segmented line. This allows you to trace the root.

    b. Trace the root, clicking when you need to change the direction of the line. When you get to the end of the root, double click to stop drawing.

    TIP: adjusting the brightness/contrast of the image can make it easier to see the roots. To do this, Click Image: Adjust: Brightness/Contrast. In the pop-up window, click Auto. This usually gives good results, but if you are not happy, you can manually adjust the levels.

    c. Click Analyze: Measure. In the Results window, review the length parameter. If you have set the scale correctly, this will be the length of the root in cm. You can only trace one root at a time, so be sure to click measure after each root. The Results window will hold all of the measurements from one image and can be saved and/or copied into Excel if desired.

    TIP: explore the other tools in ImageJ, such as the polygon tool for leaf area measurements or the angle tool for quantifying gravitropic response.

E.2) EZ-Rhizo

After downloading EZ-Rhizo, refer to the website for a tutorial, video demonstration and a forum for user feedback. Visit http://www.psrg.org.uk/plant-biometrics.html to download EZ-Rhizo.

Appendix F: Seed sterilisation and plating

This procedure is a guide for seed sterilisation and plating on MS agar.

F.1) Materials

1. 70% (v/v) ethanol (freshly prepared)
2. 70% (v/v) ethanol in spray bottle
3. bleach (commercial)
4. laminar flow or chemical hood
5. microcentrifuge
6. MS agar
7. P1000 micropipette
8. P1000 micropipette tips
9. parafilm strips (transversal cut, 2-cm wide)
10. Petri dishes
11. scale
12. scissors
13. sterile Pasteur pipette
14. sterile Petri dishes; square Petri dishes allow for sowing more seeds
15. sterile water
16. vortex
17. Tween 20
18. water bath or oven to keep melted MS agar

F.2) Solutions

Bleach/Tween solution				
25% (v/v) bleach				
0.005% (v/v) Tween 20 (5 microlitres (μl) in 1 millilitre (ml))[8]				
70% (v/v) ethanol				
0.1% (w/v) agar (in water, autoclaved)				
Sterile water (autoclaved)				
Reagent for growing medium (MS agar)	**Cat#**	**[initial]**	**[final]**	**Amount**
Murashige – Skoog salts mixture	Sigma, M5524		4.3 grams (g) L^{-1}	4.3 g

---

[8] 5 μl of Tween 20 in 1 ml of 1:4 bleach was used in Pg 2645.

MS vitamin solution	Sigma, M3900	1,000X	1X	1 ml
agar	Sigma, A1296		0.7%	7 g

Bring it up to 900 ml, bring it up to pH 5.65–5.80 with 10M KOH

F.3) Procedure

Pouring of plates

Materials	Method
Water bath or oven	Place autoclaved MS agar in 65 ºC water or 45 ºC drying oven after removing from autoclave. Use it before it solidifies.
Laminar flow	Turn on laminar flow or chemical hood. Pouring of plates and seed sterilisation must be done in a sterile environment.
70% (v/v) ethanol in spray bottle	Spray laminar flow bench and walls with 70% (v/v) ethanol and wipe with paper towel.
Petri dishes, scissors	Spray a new bag of Petri dishes with 70% (v/v) ethanol before placing into the chamber. Open the bag once inside the laminar flow.
MS agar	Place approximately 15–20 ml of the MS agar per plate, place the lid half way, and let it cool for one hour at room temperature. It is important that cooling is done with open lids to remove excess moisture. Remember that plates will be stored at 4 ºC later on and this will cause water vapor to condense, resulting in water droplets forming on the inside of the plates, preventing light diffusion and increasing chances of contamination.

Seed sterilisation using bleach solution and plating

Materials	Method
Scale, seeds, spin	Weigh approximately 20–30 milligrams (mg) of seeds and place them in 'non-sterile' 2 ml Eppendorf tubes.

70% ethanol, vortex	Add 1 ml 70% ethanol[9], vortex, incubate five minutes, spin briefly to pellet the seeds and discard supernatant.
Laminar flow	The next steps have to be performed in the laminar flow.
Sterile water, bleach	Add 750 µl of sterile water + 1 µl Tween 20 + 250 µl bleach (25% (v/v) final concentration) and incubate for **three minutes**[10].
Sterile water	Spin briefly, wash three times in 1 ml sterile water, and resuspend in 2 ml sterile 0.1% agar. You can let it decant by gravity but a quick spin (which could be done outside the laminar flow) will hasten the procedure and create a more compact seed pellet.
Sterile Pasteur pipette	Seeds are ready to be placed on the MS agar using a long-neck glass Pasteur pipette or regular pipette. NOTE: the FINAL seed density is important to dispense individual seeds and it needs to be optimised. As a guide, using 25 mg of seeds, resuspend in 2 ml sterile 0.1% agar and use a sterile glass Pasteur pipette to evenly distribute the seeds on plates.

---

[9] 70% ethanol disrupts the external membranes of bacterial cells.

[10] Bleach kills fungal spores and bacteria. Tween 20 is a detergent that breaks the surface tension of the sterilising solution, allowing better contact between the solution and the surface of the seeds.

Appendix G: Hazard sheet (risk assessment)

A risk assessment is required to be included in an experimental protocol to ensure that anyone who reads or performs the experiment understands the health risk and control measures required to work in a safe manner with the hazardous substances that may be necessary for the experiment.

The assessment involves considering the nature and severity of the potential health hazard and the route and level of exposure.

A risk assessment form must be filled in for each hazardous substance that is used in the experiment and attached to the experiment. The form requires minimal written information and covers the type of hazard, disposal, control measures and emergency first aid.

Material Safety Data Sheets (MSDS) sheets will be provided for all hazardous substances used in the Plant Detectives Practical.

Acknowledgements

The publication of this Manual was funded by an eTEXT Grant Scheme from The Australian National University (ANU). We thank all the teaching staff and students at the Research School of Biology, ANU, who helped in the developing this Manual over the years. In particular, we appreciate the input from Professors Susanne von Caemmerer and Marilyn Ball for their contributions as lecturers and to the practical session. Dr Elizabeth Beckman held regular feedback sessions with students and introduced pedagogic principles we applied during the practical. Dr Terry Neeman provided expertise in the statistical analysis. Dr Amy Davidson is acknowledged for her efforts to help establishing the class resources and her ideas for quiz design and implementation. We are also thankful to Dr Verónica Albrecht for convening the class and commenting on the Manual drafts. This Manual was also highly improved by the revision of earlier versions by former students Rodney Eyles, John Rivers, Estee Tee, Nadiatul Mohd Radzman and Christina Delay. Erin Walsh and Christina Delay provided detailed instructions on how to analyse images using ImageJ and Photoshop, respectively, and Nadiatul Mohd Radzman also contributed with the "Before the practical component" session. We also thank all the peer mentors and students over the years for their hard work and the invaluable feedback they provide.

Bibliography

Armengaud P, Zambaux K, Hills A, Sulpice R, Pattison RJ, Blatt MR, Amtmann A (2009) EZ-Rhizo: integrated software for the fast and accurate measurement of root system architecture. *The Plant Journal* 57 (5):945–56

Arvidsson S, Pérez-Rodríguez P, Mueller-Roeber B (2011) A growth phenotyping pipeline for *Arabidopsis thaliana* integrating image analysis and rosette area modeling for robust quantification of genotype effects. *New Phytologist* 191 (3):895–907. doi:10.1111/j.1469-8137.2011.03756.x

Bower JE, Zar JH, von Ende CN (1989) *Field and Laboratory Methods for General Ecology*. Wm. C. Brown Company Publishers, Dubouque, Iowa, USA

Boyes DC, Zayed AM, Ascenzi R, McCaskill AJ, Hoffman NE, Davis KR, Gorlach J (2001) Growth stage-based phenotypic analysis of *Arabidopsis*: A model for high throughput functional genomics in plants. *Plant Cell* 13 (7):1499–510

Clark, S. E. (2001) Cell signalling at the shoot meristem. *Nat Rev Mol Cell Biol* 2(4): 276-284.

Desikan R, Griffiths R, Hancock J, Neill S (2002) A new role for an old enzyme: Nitrate reductase-mediated nitric oxide generation is required for abscisic acid-induced stomatal closure in *Arabidopsis thaliana*. *Proc Natl Acad Sci USA* 99 (25):16314–18

Förster B, Osmond CB, Pogson BJ (2009) De novo synthesis and degradation of Lx and V cycle pigments during shade and sun acclimation in avocado leaves. *Plant Physiology* 149 (2):1179–95. doi:10.1104/pp.108.131417

Hoffmann WA, Poorter H (2002) Avoiding bias in calculations of relative growth rate. *Annals of Botany* 90 (1):37–42. doi:10.1093/aob/mcf140

Hunt R, Causton DR, Shipley B, Askew AP (2002) A modern tool for classical plant growth analysis. *Annals of Botany* 90:485–88

Initiative TAG (2000) Analysis of the genome sequence of the flowering plant *Arabidopsis thaliana*. *Nature* 408 (6814):796–815. doi: www.nature.com/nature/journal/v408/n6814/suppinfo/408796a0_S1.html

Jones HG (2007) Monitoring plant and soil water status: Established and novel methods revisited and their relevance to studies of drought tolerance. *Journal of Experimental Botany* 58:119–30

Lavagi I, Estelle M, Weckwerth W, Beynon J, Bastow RM (2012) From bench to bountiful harvests: A road map for the next decade of *Arabidopsis* research. *The Plant Cell Online* 24 (6):2240–47. doi:10.1105/tpc.112.096982

Lee J, Durst RW, Wrolstad RE (2005) Determination of total monomeric anthocyanin pigment content of fruit juices, beverages, natural colorants, and wines by the pH differential method: Collaborative study. *Journal of AOAC International* 88 (5):1269

Lee J, Rennaker C, Wrolstad RE (2008) Correlation of two anthocyanin quantification methods: HPLC and spectrophotometric methods. *Food Chemistry* 110 (3):782–86

Meinke DW, Cherry JM, Dean C, Rounsley SD, Koornneef M (1998) *Arabidopsis thaliana*: A model plant for genome analysis. *Science* 282 (5389):662–82

Monteith JL, Campbell GS, Potter EA (1988) Theory and performance of a dynamic diffusion porometer. *Agricultural and Forest Meteorology* 44 (1):27–38. doi:dx.doi.org/10.1016/0168-1923(88)90031-7

Morita MT, Tasaka M (2004) Gravity sensing and signaling. *Current Opinion in Plant Biology* 7 (6):712–18. doi:dx.doi.org/10.1016/j.pbi.2004.09.001

Muller K, Tintelnot S, Leubner-Metzger G (2006) Endosperm-limited *Brassicaceae* seed germination: Abscisic acid ihibits embryo-induced endosperm weakening of *Lepidium sativum* (cress) and endosperm rupture of cress and *Arabidopsis thaliana. Plant and Cell Physiology* 47 (7):864–77. doi:10.1093/pcp/pcj059

Neff MM, Chory J (1998) Genetic interactions between phytochrome A, phytochrome B, and cryptochrome 1 during Arabidopsis development. *Plant Physiology* 118 (1):27–35

Roelfsema MRG, Prins HBA (1995) Effect of abscisic acid on stomatal opening in isolated epidermal strips of abi mutants of *Arabidopsis thaliana. Physiologia Plantarum* 95 (3):373–78

Turner, S. R. and C. R. Somerville (1997) Collapsed xylem phenotype of Arabidopsis identifies mutants deficient in cellulose deposition in the secondary cell wall. *The Plant Cell* 9 (5): 689-701.

United Nations Food and Agriculture Orgnanisation, http://www.fao.org/fileadmin/templates/wsfs/docs/expert_paper/How_to_Feed_the_World_in_2050.pdf

Weigel D (2012) Natural Variation in Arabidopsis: From Molecular Genetics to Ecological Genomics. *Plant Physiology* 158 (1):2–22. doi:10.1104/pp.111.189845

i World Hunger Education Service (2012) *2012 World Hunger and Poverty Facts and Statistics*.
http://www.worldhunger.org/articles/Learn/world%20hunger%20facts%202002.htm

ii Croplife (2012) Submission in Response to National Food Plan Green Paper. Introduction, paragraph 4.
http://www.croplifeaustralia.org.au/files/newsinfo/submissions/2012/CropLife%20Submission-National%20Food%20Plan.pdf

www.ingramcontent.com/pod-product-compliance
Lightning Source LLC
Chambersburg PA
CBHW051308270326
41928CB00027B/3453